Management and Industrial Engineering

Series Editor

J. Paulo Davim, Department of Mechanical Engineering, University of Aveiro, Aveiro, Portugal

This series fosters information exchange and discussion on management and industrial engineering and related aspects, namely global management, organizational development and change, strategic management, lean production, performance management, production management, quality engineering, maintenance management, productivity improvement, materials management, human resource management, workforce behavior, innovation and change, technological and organizational flexibility, self-directed work teams, knowledge management, organizational learning, learning organizations, entrepreneurship, sustainable management, etc. The series provides discussion and the exchange of information on principles, strategies, models, techniques, methodologies and applications of management and industrial engineering in the field of the different types of organizational activities. It aims to communicate the latest developments and thinking in what concerns the latest research activity relating to new organizational challenges and changes world-wide. Contributions to this book series are welcome on all subjects related with management and industrial engineering. To submit a proposal or request further information, please contact Professor J. Paulo Davim, Book Series Editor, pdavim@ua.pt

More information about this series at http://www.springer.com/series/11690

Carolina Machado · J. Paulo Davim
Editors

Coaching for Managers and Engineers

Springer

Editors
Carolina Machado
School of Economics and Management
University of Minho
Braga, Portugal

J. Paulo Davim
Department of Mechanical Engineering
University of Aveiro
Aveiro, Portugal

ISSN 2365-0532 ISSN 2365-0540 (electronic)
Management and Industrial Engineering
ISBN 978-3-030-71107-8 ISBN 978-3-030-71105-4 (eBook)
https://doi.org/10.1007/978-3-030-71105-4

This Springer imprint is published by the registered company Springer Nature Switzerland AG
The registered company address is: Gewerbestrasse 11, 6330 Cham, Switzerland

Preface

Defined by the International Coach Federation as a continuous shared process that allows the client to obtain satisfying results both in his personal and professional life; a process through which the client deepen his knowledge, improve his performance as well as his life quality, the coaching construct is present nowadays in all competitive and dynamic organizations. Indeed, not saying what to do, the coaching focuses its attention on helping professionals to clarify and control their lives. In other words, it helps these professionals to evaluate what they are doing in their lives, taking into account their goals, values, dreams, desires and intentions. It doesn't give answers. On the contrary, it puts questions, shows new options and leads to the change. It is a relationship between the coach and the client; a process whereby is provided feedback and collaborative dialogue with the client with a view to its growth.

Given its characteristics, the coaching process is a critical tool that nowadays managers and engineers need in order to better find, obtain and make decisions which best cater for their and the organization interests.

Conscious of its importance in nowadays organizations, the present book entitled *Coaching for managers and engineers* looks to communicate the latest developments and thinking on the coaching subject worldwide. This is a critical tool to managers and engineers as it makes easier that these professionals can face and adapt to the organizational changes in a more efficient and effective way. At the same time that focusing on the main values and commitment of the human being helps them to find themselves, to be responsible for their life, renewing and favouring communication in relationships. Highlighting these professionals' potentialities, *Coaching for Managers and Engineers* allows them to reach aims usually seen as impossible to reach.

Seeking critical inputs from studies related to coaching issues, in the present book, we can find diverse contributions, exceptional not only in terms of theory but also in practical and/or methodological terms and that significantly advance social scientific research on the field of coaching in organizations. This book is designed to increase the knowledge and effectiveness of all those involved in these areas whether in the profit or non-profit sectors or in the public or private sectors.

Organized in six chapters, *Coaching for Managers and Engineers* discusses in chapter "From Theory to Practice—in Search of Theoretical Approaches

Leading to Informed Coaching Practices" while second chapter presents "Building Blocks of Coaching Project Managers". Chapter 3 deals with "Socialization Coaching: An Application for Welcoming and Integrating People ("The Point Out of the Curve") Without Homogenizing"; chapter 4 speaks about "Coaching for All—New Approaches for Future Challenges". Chapter 5 focuses "Prone to Follow, Eager to Lead: Millennials as the Ultimate Commodity on the Job Market". Finally, chapter 6 looks to cover "The Dark Side of Human Resources Management: The Perceptions of Different Organizational Actors".

Particularly oriented to managers and engineers, *Coaching for Managers and Engineers* can also be used for academics, researchers and other professionals that are interested in developing coaching processes in their personal lives and/or in the organization.

Given the high potentialities of coaching for managers and engineers, the interest in this subject, at present, is evident for many types of organizations, namely, important institutes and universities in the world.

The editors acknowledge their gratitude to Springer for this opportunity and for their professional support. Finally, we would like to thank all chapter authors for their interest and availability to work on this project.

Braga, Portugal Carolina Machado
Aveiro, Portugal J. Paulo Davim

Contents

About the Editors

Carolina Machado received her Ph.D. degree in Management Sciences (Organizational and Politics Management area/Human Resource Management) from the University of Minho in 1999, Master degree in Management (Strategic Human Resource Management) from the Technical University of Lisbon in 1994 and Degree in Business Administration from the University of Minho in 1989. She is teaching Human Resource Management subjects since 1989 at the University of Minho; she is working as an Associated Professor since 2004, with experience and research interest areas in the field of Human Resource Management, International Human Resource Management, Human Resource Management in SMEs, Training and Development, Emotional Intelligence, Management Change, Knowledge Management and Management/HRM in the Digital Age/Business Analytics. She is the Head of the Human Resource Management Work Group at the School of Economics and Management of the University of Minho, Coordinator of Advanced Training Courses at the Interdisciplinary Centre of Social Sciences, Member of the Interdisciplinary Centre of Social Sciences (CICS.NOVA.UMinho), University of Minho, as well as Chief Editor of the International Journal of Applied Management Sciences and Engineering (IJAMSE), Guest Editor of journals, books Editor and Book Series Editor, as well as a reviewer for different international prestigious journals. In addition, she has also published both as editor/co-editor and as author/co-author several books, book chapters and articles in journals and conferences.

J. Paulo Davim is a Full Professor at the University of Aveiro, Portugal. He is also distinguished as honorary professor in several universities/colleges/institutes in China, India and Spain. He received his Ph.D. degree in Mechanical Engineering in 1997, M.Sc. degree in Mechanical Engineering (materials and manufacturing processes) in 1991, Mechanical Engineering degree (5 years) in 1986, from the University of Porto (FEUP), the Aggregate title (Full Habilitation) from the University of Coimbra in 2005 and the D.Sc. (Higher Doctorate) from London Metropolitan University in 2013. He is a Senior Chartered Engineer at the Portuguese Institution of Engineers with an MBA and Specialist titles in Engineering and Industrial Management as well as in Metrology. He is also an Eur Ing by FEANI, Brussels and a Fellow of IET (FIET), London. He has more than 30 years of teaching and research experience in Manufacturing, Materials, Mechanical and Industrial Engineering, with special emphasis on Machining and Tribology. He also has interest in Management, Engineering Education and Higher Education for Sustainability. He has guided large numbers of postdoc, Ph.D. and master's students as well as has coordinated and participated in several financed research projects. He has received several scientific awards and honours. He has worked as an evaluator of projects for European Research Council (ERC) and other international research agencies as well as examiner of Ph.D. thesis for many universities in different countries. He is the Editor in Chief of several international journals, Guest Editor of journals, book Editor, book Series Editor and part of Scientific Advisory for many international journals and conferences. Presently, he is an Editorial Board member of 30 international journals and acts as reviewer for more than 100 prestigious Web of Science journals. In addition, he has also published as an editor (and co-editor) more than 200 books and as author (and co-author) more than 15 books, 100 book chapters and 500 articles in journals and conferences (more than 280 articles in journals indexed in Web of Science core collection/h-index 59+/11000+ citations, SCOPUS/h-index 63+/14000+ citations, Google Scholar/h-index 81+/23000+ citations). He is listed in the World's Top 2% Scientists by Stanford University study.

From Theory to Practice—in Search of Theoretical Approaches Leading to Informed Coaching Practices

Carla Gomes da Costa and Andrea Fontes

Abstract In order to define Coaching, we first examined what Coaching is and what it is not. Then, we compared Coaching with other types of interventions. We also considered the different types of Coaching. Being aware of all the challenges that Coaching must deal with, it is also our aim in this chapter to stress those approaches that are most relevant to orienting best practices, particularly in the organisational context.

1 Introduction

Determining a single, complete or intemporal definition of Coaching is a challenging enterprise that must involve the identification of several variables. From scientific research into this knowledge domain, it is possible to identify the most common variables that explain the difference between what Coaching is and what it is not since these concern the big theoretical-practical Coaching issues. In the main, these variables are the purpose of Coaching, the different types of Coaching, the context in which Coaching takes place, and the results expected from the Coaching process.

The *purpose of Coaching* is related to instilling motivation and responsibility in the Coachee in order for them to design and accomplish an action plan. This action plan will lead to improvement and change, specifically to improved performance in a professional domain and to behavioural change, either in a professional or personal context. Training and Consultancy can also promote this, but these are not the core goals of those specific interventions. The same can be said of Psychotherapy and

The original version of this chapter was revised: Belated correction has been incorporated. The correction to this chapter is available at https://doi.org/10.1007/978-3-030-71105-4_7

C. G. da Costa (✉)
Instituto Superior Manuel Teixeira Gomes, Lisbon, Portugal
e-mail: carlagomescosta@ismat.pt

A. Fontes
ISCTE, Instituto Universitário de Lisboa, Lisbon, Portugal

C. Machado and J. P. Davim (eds.), *Coaching for Managers and Engineers*, Management and Industrial Engineering,
https://doi.org/10.1007/978-3-030-71105-4_1

Counselling, which could indirectly promote these goals, but that is not their main purpose either. The specificity of Coaching must highlight the non-clinical population as the target Clients.

Setting aside the existing classifications, *types of Coaching* can be mainly divided into the Coaching that is done within organisations through performance orientation, and outside organisations through other global orientation. The Coaching process is deeply related to the *context* in which it is activated. In modern societies, the major context in which Coaching has been developed is the work context where the purpose is deeply related to performance improvement. This use of Coaching is goal-oriented, that is to say, it is done to achieve e*xpected results,* such as to improve skills and competencies and specific behaviours that can be identified and measured, if possible, by performance indicators.

This discussion would not be complete without mentioning the different approaches that have influenced the practice of Coaching, and that involves the way the r*elationship setting* is managed.

It being essential in helping provide rapid adaptation to challenges, Coaching has a special status in society today and, as such, the professional training of Coaches must be paramount for Coaching Associations or Coaching Societies in each country.

2 Coaching—to Be or not to Be…in Search of a Possible Definition for Coaching…

Despite the plethora of information about Coaching, people in general know little about the Do's and Don'ts regarding branches, processes, goals, and results. There is a tendency to believe that when something new comes along, it is a magic wand that with one wave will facilitate change processes and enhance people's well-being. Such is the case with coaching, where the multifarious information available can be very confusing and lead to misunderstandings. Coaching can stem from a variety of theoretical approaches, each of which will condition specific practice interventions and involve a multitude of techniques (Grover and Furnham 2016).

The dilemma is whether it is easier to be a Coach, or in the Coaching process as a Coachee, than to present a complete and perfect definition of Coaching. In a scientific sense, to say that something is difficult to define, when everyone can do it with little or no previous (or proper) preparation makes it sound more like an art or, even worse, something more akin to selling snake oil. Nowadays, there are mixed feelings with regard to Coaching due to its being such a widespread big business that gleans millions without any concrete proof of its efficacy. The various criticisms about the principles of Coaching and the nonexistence of a common rationale among Coaching experts regarding practice (and also between researchers of different branches of Coaching) stems from a lack of any deep conceptual framework able to shed some light on all this.

At the heart of Coaching principles are the main characters, their roles and the rapport they establish between them. There must be a Coach and at least one Coachee (assuming individual interventions) and their roles are predefined. The Coach is responsible for understanding the Coachee's frame of reference and their objectives. The Coach needs to create the conditions in which the Coachee can discover their resources, identify obstacles and develop strategies to overcome them, and define efficient actions towards achieving their goal. The Coachee should be responsible for their own development, define their objectives, intermediate targets and action plan, and provide the resources to achieve their goals.

Setting aside a single definition and an exclusive type of theoretical perspective, there are nevertheless some common aspects that can be identified in the Coaching literature (Connor and Pokora 2007; Cox et al. 2009; Smith and Brumell 2013) especially with regard to non-directive approaches, and in organisational contexts. These common aspects are basic Coaching principles that function as boundaries to the practice. Among them are:

- *Coaching as a developmental instrument*—Coaching is an instrument used to develop competencies in another (Coachee), leading to performance improvements.
- *Coaching is done by a professional*—it is possible to identify the difference between being a professional or having a Coaching attitude in the way we talk to people. A Coach facilitates the process, they are a facilitator and not a magician that will conjure solutions.
- *Coaching is to respect Coachees' agenda*—it is accepted now that to get more results a Coachee needs to be involved in their own goals in line with organisational goals.
- *Coaching is not to be confused with other developmental or health-related practices*—as a specific practice, Coaching is not the same as Mentoring, Training, Consultancy, Counselling, Psychotherapy.
- *Coaching is not to be used with other practices*—even if you are a Psychotherapist, a Psychologist, a Trainer, if you are acting as a professional Coach you must be aware of the limits to intervention that Coaching involves.
- *Coaching involves a collaborative relationship*—as an alliance between Coach and Coachee, it is based on confidentiality, action steps are designed to better achieve desired goals, and expected results, and to keep the Coachee motivated.
- *Coaching goal achievement as a Coachee's responsibility*—the Coach is a facilitator, but the action involvement and goal achievement are the responsibility of the Coachee.
- *Coaching uses communication techniques to facilitate Coachees' awareness*—Being empathetic, knowing how to formulate questions and active listening are communication skills that activate Coachees' awareness and facilitate the process of defining the Coaching steps.

When the above principles are respected in Coaching practice, it is easy to understand the powerful role a motivational relationship can play with regard to building the special bond that leads to the commitment that is the basis for success in the Coaching process. For the Coachee, a supportive relationship also helps to reduce stress, and facilitates the improvement of psychological resources, such as self-awareness; self-efficacy; resilience and problem-solving, among others.

Awareness of some inner and basic Coaching principles makes it easier to understand what the common aspects are able to identify and to view Coaching as an autonomous and specific intervention. Above all, and setting aside the different theoretical approaches, coaching is:

- An independent, specific technique, different from Mentoring, Training, Consultancy, Counselling, Psychotherapy.
- A relational professional practice, where the roles and responsibilities of the Coach and the Coachee are defined.
- An intervention that follows systematic steps, based on a specific theoretical approach.
- Focused on previously defined goals and on achieving them.
- An intervention that can be evaluated by comparing the results with the previously defined goals.

Coaching is a widespread term currently used to describe several types of interventions which do not, however, all lead to effective Coaching. It is, therefore, important to clarify which characteristics of Coaching are to be considered key to Coaching interventions and, consequently, which are to be excluded.

3 Coaching Comparison with Other Types of Intervention

It is also important to differentiate Coaching from other practices that could have some similarities, such as Counselling or Psychotherapy. Regarding Counselling and Psychotherapy, the main difference resides in the temporal moment the sessions focus on. While those types of intervention usually delve into a person's past history in order to uncover the origins of their current way of thinking and behaviours, Coaching on the other hand focuses on the future, and develops strategies to pursue predetermined goals (Theeboom et al. 2014). What is more, the Coaching relationship should be a non-therapeutic one (Grant and O'Connor 2010). Indeed, if an individual has a limited capacity for change, it could be very difficult and even counterproductive for them to embark upon a Coaching program (McKenna and Davis 2009).

Taking an organisational context, and assuming there is a form of Coaching specifically designed to improve skills, competencies and performance within organisations (Fontes and Dello Russo 2019; Hamlin et al. 2008), it is important to compare that with other forms of in-house intervention such as Mentoring, Consulting and Training. In the case of Mentoring, it is assumed that the Mentor must have more experience in either the job itself or in the company culture (Dello Russo et al. 2016). In Coaching, any assumed superior knowledge of the coach does not exist, and there does not even need to be a shared professional background, although it has been acknowledged that it can be helpful towards guaranteeing a better understanding of the Coachee's context (Bono et al. 2009).

The differences between Coaching and Consulting are mainly in the output of the one providing the service: the Consultant gives specific recommendations about the best way to achieve a predefined objective or solve a problem; the Coach should not give opinions or provide solutions but should pose the questions that will lead the Coachee to find their own answers. We found that Training also differs from Coaching with regard to output, with the difference being in the way it is presented. In Training, the Trainer is expected to employ content previously agreed with the organisation. In Coaching, however, there is no agreed, predefined content since the content to be discussed will be brought by the Coachee, and not the Coach. To reiterate then, in Coaching, unlike in both training and Consulting, the Coach is not expected to provide any content or make recommendations but should help the Coachee, usually through questioning, to find their own answers and solutions.

4 Types of Coaching

As previously mentioned, Coaching is a broad term that was first used in association with sports but its application then extended to several other areas. In Table 1, several types of Coaching are shown according to three main criteria: the topic being discussed, the Coach and to whom the service is provided, and the Coachee.

Our focus will be on organisational Coaching, which means the Coaching provided in a corporate context. Since it is not unusual to find it being called different names, such as executive or business, we shall differentiate between types depending on who is receiving the Coaching: when provided to the top management we refer to it as Executive Coaching, when it is available to those who manage teams we call it Leadership Coaching, and when it is available to all employees we call it Business Coaching.

The Coach can have different positions on the organisational chart of a company. They might be included in the company structure, and considered an internal Coach or, contrastingly, be considered an external service provider not related to the

Table 1 Somes types of Coaching

Topic	Non-organisational: Life; Career; Sports; Health &Wellbeing; Educational; Parental,…	Organisational
Coach		Internal; External; *Team*; Managerial*; Human Resources*; Peer**
Coachee		Business; Leadership; Executive
Form	Face to face; telephone; Online; Blended; Individual; Group	

*Not considered Coaching by the majority of scholars

company, and hence designated an external Coach. There are advantages and disadvantages to both types. An internal Coach knows the culture of the company in-depth, which can facilitate the Coaching process, but it can also be restrictive depending on the boundaries of the employing organisation. An external Coach would not necessarily be limited by any such boundaries (Jones et al. 2016).

We believe it is important to mention other forms that could be called Coaching such as peer, team, managerial Coaching or even Coaching provided by Human Resources. In our understanding, those forms of intervention do not comply with one of the important prerequisites of Coaching, namely the unbiased role of the Coach. In all these other forms, the Coach has a specific role in the company and their own goals. For example, imagine the case of a Coachee whose goal is to be promoted, or to be transferred to another department: openly discussing that goal at an early stage and seeking strategies to achieve it, with a peer, the manager or even Human Resources could create some early controversy and possibly become a stumbling block in the Coaching process. That is why we consider that those kinds of interactions are closer to Mentoring than to Coaching. The Coaches in that case (peers, managers, or human resources professionals) can, in fact, use their Coaching knowledge and skills, but it must be borne in mind that the role they occupy within the organisation might impede the "pure" coaching process.

Last but not least, it is important to note that organisational Coaching has one particular specificity, namely a third party additional to the Coach and the Coachee and which is normally the entity that finances the Coaching program: the company. It is important from the very beginning to establish the limits of the role of each intervenient, as well as to get an agreement of confidentiality that protects the relationship between Coach and Coachee. Unfortunately, it is not uncommon to come across obstacles that can present challenges to the Coaching practice in an organisational setting.

Reflection moment—Exercise 1

Considering what you have read above, how would you answer the following:

1. If the Coaching service is paid for by the company, can the company limit the type of goals to those related to the performance of the Employer/Coachee?

   ____

   ____

2. Conversely, if the Coachee brings a topic from his private life to the Coaching session, what should the Coach do?

   ____

   ____

3. If the Coachee brings to the session his intention to leave the company and asks the Coach to guide him to the best decision, what should the Coachee do?

____

____

4. If the CEO approaches the Coach and mentions something about a certain Coachee being at risk of being fired due to poor performance, what should the Coach do with that information in the next session with the Coachee?

____

____

5 Different Coaching Approaches

The difficulties found in defining Coaching are compounded when it comes to dealing with identifying possible Coaching approaches and their relationship to practice. To achieve scientific status as a systematic intervention and gain more credibility at a social level, each Coaching intervention must be identified with some theoretical approaches which can then inform practical interventions with Clients.

Coaching theory is grounded in several areas, namely: psychology, such as positive psychology (Seligman and Csikszentmihalyi 2000); and humanistic, constructivist and coaching psychology (Passmore 2010). Neurolinguistic Programming (NLP), is also commonly associated with Coaching. NLP, as explicit in the name, merges neurology, language and programming. Its multidimensional processes include the development of behavioural competence and flexibility, strategic thinking and cognitive processing. It employs several techniques whose origins are in linguistics, psychology, systems theory, cybernetics, and hypnosis (Fontes and Dello Russo 2019). However, being derived from the human and social sciences and psychotherapy schools, this very diversity of background could be either a strength for its wealth of knowledge and instruments or a weakness through misuse of techniques unassociated to specific theoretical approaches.

To understand the practices of Coaching, it is relevant to take into consideration the dimensions put forward by Yves (2008) that characterise Coaching approaches. To begin with, the practices could be considered as being directive or non-directive in the way responsibilities for the process are attributed more to Coach or the Coachee. Second, Coaching could also be focused more on personal development or be more goal-focused. Goal-oriented Coaching as a specific paradigm comprises several characteristics: non-directivity, goal-oriented and performance-driven. And third, Coaching could be derived from therapeutic approaches or performance-driven intentions.

At its inception, Coaching was seen as a partially directive intervention, with the Coach being considered as someone who could direct a Coachee to find a solution (Yves 2008). From being a partially directive intervention, Coaching has evolved to being more of a non-directive intervention, where the responsibility is assumed by Coachee.

In the interest of openness and out of respect for all theoretical perspectives and forms of intervention, it is relevant to mention just a few approaches (Stober and Grant 2010) that have influenced and oriented the way Coaching has built its role as a performance improvement tool within the organisational context.

The Humanistic Perspective based on the Maslow (1954) legacy and also on the Rogerian (Rogers 1959) principles of person-centred theory, developed and spread a theory that the inner individual motivation is toward positive change and self-actualisation. This perspective has had a very positive influence on valuing human development potential within organisations, making room for Coaching to assert itself as an intervention that can facilitate it. More recently, this humanistic perspective has been associated with other theories such as the Self-Determination Theory (SDT; Gregory and Levy 2012). According to their model, coaching practices (autonomy-, competency- and relatedness-supportive behaviours) influence employee outcomes through the attainment of psychological needs: autonomy, competence and relatedness (Gabriel et al. 2014).

The Behavioural Approach (Eldridge and Dembkowski 2013) is based on practical behaviour change and highlights the need to achieve a better adjustment to work demands by changing specific behaviours in accordance with defined goals.

The Cognitive Approach (Ellison and Hayes 2013) has pointed out the maladaptive beliefs that informed inaccurate cognitions and influenced emotions and feelings, explaining a maladjustment to reality. This approach has influenced Coaching by challenging Coachees' distorted perceptions in order to facilitate overcoming them and develop a more adaptative way to read reality.

The Positive Psychology Approach (Seligman and Csikszentmihalyi 2000) has stressed that a positive personal disposition improves happiness. Coaching, by reinforcing a positive psychological disposition, facilitates performance improvement. By focusing on the positive aspects of their lives Coachees' can more easily access the psychological resources responsible for improving performance.

Finally, the Goal-Oriented Approach (David et al. 2014) is strictly centred on the relevance of Coaching as a tool that directs individuals to be oriented and motivated towards goal achievement.

However, there are many other theoretical approaches associated with organisational Coaching. Some of them are presented in Table 2 in this adaptation of what Cox et al. (2009) included in their book. In keeping with the authors' intention, this table aims to illustrate the variety of influences on Coaching practice, and some others that might also be found.

Table 2 Coaching in Organisations and theoretical traditions (adapted from Cox et al. (2009))

	Skills & Performance Coaching	Developmental Coaching	Executive & Leadership Coaching	Career Coaching	Team Coaching
The psychodynamic approach to coaching	*	*	**		*
Cognitive-behavioural coaching	**	**	**	**	**
The solution-focused approach to coaching	**	**	**		*
The person-centred approach to coaching	*	**	*	**	*
The Gestalt approach to coaching	*	*	*	*	**
Existential coaching	*	*	*	*	*
The transpersonal approach to coaching	*	*	**	*	**
Positive psychology approach to coaching	**	*	*	*	*
Transactional analysis and coaching	*	*	*	*	*
The NLP approach to coaching	**	*	*	*	*

Note The stars in the matrix indicate that a link is made, whether to a particular theoretical chapter on the horizontal dimension or by the practitioners or a context from the vertical dimension

6 Outcomes of Coaching

As could be expected from the theoretical implications, the outcomes of coaching show a multiplicity of benefits associated with its practice. With our focus exclusively on the organisational context, we found two relevant meta-analytic studies (Jones et al. 2016; Theeboom et al. 2014). Following the classification used by Jones and colleagues (2016) the main outcomes can be organised into categories like affective, cognitive and skill-based. In the affective category, attitudinal and motivational outcomes such as self-efficacy, self, confidence, job satisfaction and organisational commitment are included. In the skill-based outcomes, we can find several competencies, such as leadership skills, communication skills and other individualised technical skills. Finally, in the cognitive outcomes, we find that the development of new cognitive strategies can take several forms, among them being problem-solving and solution-focused competencies. What is more, several other outcomes can be found in the workplace environment, such as those related to well-being at work, like reduced instances of depression and burnout, or others related to self-regulated actions, like goal setting and goal attainment or improved performance overall (Theeboom et al. 2014).

Exercise 2—Brief exercise rationale-setting goals
Setting goals is the most important initial task in coaching. A clear definition of goals facilitates their achievement. To establish goals in life and in your career has great motivational potential since the goals defined are bespoke and should relate to one's most important values, main interests and motivational profile. Being the author of one's own goals and being aware of why they are important for you, makes you feel responsible and involved in their attainment. This is why it is crucial that the agendas of the Coachee and the Organisation must match. If a company wants to help you to develop leadership competencies, but you do not see yourself as a leader, it will not work. Goal setting must not be a Coach/Organisation imposition, but the result of a negotiation process related to your career intentions. Sometimes, it is difficult to find this balance, but the success of the Coaching process depends on it.

Several categorisations have been developed to help people establish clearly defined goals. The SMART objectives categorisation is one of the most commonly known. We invite you to reflect on specific goals that could apply in your workplace context.

Reflection moment—Exercise 2—Defining goals

Considering your professional area, define three goals related to what you want to achieve and write them in the space provided below.

To help refine your definition please consider the following categorisations and their specific characteristics:

SMART	Specific Measurable Tangible Realistic Timing
PURE	Positive Understood Relevant Ethical
CLEAR	Challenging Legal Ecologic

References

Bono, J. E., Purvanova, R. K., Towler, A. J., & Peterson, D. B. (2009). A survey of executive coaching practices. *Personnel Psychology, 62,* 361–404.

Connor, M., & Pokora, J. (2007). *Coaching and mentoring at work: Developing effective practice.* Buckingham: Open University Press.

Cox, E., Bachkirova, T., & Clutterbuck, D. (2009). *The complete handbook of coaching* (3rd ed.). London: SAGE Publications Ltd.

David, S., Clutterbuck, D., & Megginson, D. (2014). Goal orientation in coaching differs according to region, experience, and education. *International Journal of Evidence Based Coaching and Mentoring, 12*(2), 134.

Dello Russo, S., Miraglia, M., & Borgogni, L. (2016). Reducing organizational politics in performance appraisal: The role of coaching leaders for age-diverse employees. *Human Resource Management, 56*(5), 769–783.

Eldridge, F., & Dembkowski, S. (2013). Behavioral coaching. In: *The Wiley-Blackwell handbook of the psychology of coaching and mentoring*, 298–318.

Ellison, J. L., & Hayes, C. (Eds.). (2013). *Cognitive coaching: Weaving threads of learning and change into the culture of an organization.* Rowman & Littlefield.

Fontes, A., & Dello Russo, S. (2019). Quo Vadis? A study of the state and development of coaching in Portugal. *International Journal of Training and Development, 23*(4), 291–312.

Gabriel, A., Moran, C., & Gregory J. (2014). How can humanistic coaching affect employee well-being and performance? An application of self-determination theory. *Coaching: An International Journal of Theory Research and Practice* 7, 56–73.

Grant, A. M., & O'Connor, S. A. (2010). The differential effects of solution-focused and problem-focused coaching questions: A pilot study with implications for practice. *Industrial and Commercial Training, 42,* 102–111.

Gregory, J. B., & Levy, P. E. (2012). Humanistic/person-centered approaches. In J. Passmor, D. Peterson, & T. Freire (Eds.), *The handbook of the psychology of coaching and mentoring.* Hoboken, NJ: Wiley-Blackwell.

Grover, S., & Furnham, A. (2016). Coaching as a developmental intervention in organisations: A systematic review of its effectiveness and the mechanisms underlying it. *PLoS ONE, 11*(7), 1–41.

Hamlin, R. G., Ellinger, A. D., & Beattie, R. S. (2008). The emergent'coaching industry': A wake-up call for HRD professionals. *Human Resource Development International, 11*(3), 287–305.

Jones, R. J., Woods, S. A., & Guillaume, Y. R. (2016). 'The effectiveness of workplace coaching: A meta-analysis of learning and performance outcomes from coaching. *Journal of Occupational and Organizational Psychology, 89*(2), 249–277.

Maslow, A. H. (1954). *Motivation and personality.* New York: Harper.

McKenna, D. D., & Davis, S. L. (2009). Hidden in plain sight: The active ingredients of executive coaching. *Industrial and Organizational Psychology, 2*(3), 244–260.

Passmore, J. (2010). 'A grounded theory study of the coachee experience: The implications for training and practice in coaching psychology. *International Coaching Psychology Review, 5*(1), 48–62.

Rogers, C. R. (1959). A theory of therapy, personality, and interpersonal relationships as developed in the client-centered framework. In S. Koch (Ed.), *Psychology: A study of a science* (Vol. 3, pp. 184–256)., Formulations of the Person and the Social Context New York: McGraw-Hill.

Seligman, M. E. P., & Csikszentmihalyi, M. (2000). Positive psychology: An introduction. *American Psychologist, 55*(1), 5–14.

Smith, I. M., & Brummel, B. J. (2013). Investigating the role of the active ingredients in executive coaching. *Coaching: An International Journal of Theory, Research and Practice, 6,* 57–71.

Stober, D. R., & Grant, A. M. (Eds.). (2010). *Evidence based coaching handbook: Putting best practices to work for your clients.* John Wiley & Sons.

Theeboom, T., Beersma, B., & Van Vianen, A. E. (2014). Does coaching work? A meta-analysis on the effects of coaching on individual level outcomes in an organizational context. *Journal of Positive Psychology, 9*(1), 1–18.

Yves, Y. (2008). What is 'coaching'? An exploration of conflicting paradigms. *International Journal of Evidence Based Coaching and Mentoring, 6*(2), 100–113.

Building Blocks of Coaching Project Managers

Davar Rezania and Thomas Sasso

Abstract In this paper, we focus on coaching of project managers. We explore the quadrangular relationship between the organization, coach, project manager, and project team, and suggest that coaching project managers should include team coaching. We examine the fundamental difference between the coach–project manager and project manager–team relationships and its impact on the process of coaching.

1 Introduction

Projects are major vehicles for creating new products or services and are essential for engineering firms to reach their strategic objectives. Project management concerns the application of knowledge, skills, tools, resources, and techniques in project activities to meet the project requirements (Project Management Institute). Today, virtually all organizations are using projects as a way of organizing work and there is evidence in support of the contribution of HR strategies that focus on the development of people and teams to enhance the success of organizations (Guest 1997; Delantey and Huselid 1996). In line with this, organizational management is tasked to implement strategies to ensure that project managers achieve their full potential. Part of this strategy is putting corporate mechanisms in place that supports, promotes, and harnesses project manager leadership development (Thomas and Mengel 2008).

One mechanism that can be used in project-based organizations to develop strong leaders is a leadership coaching. Leadership coaching is often viewed as a formal relationship between a coach and a coachee aimed at developing the coachee, understanding their needs, and supporting them to reach their goals (Ely et al. 2010; Boyatzis et al. 2006). In particular, Ely et al. 2010 defined leadership coaching as a "relationship in which a client engages with a coach in order to facilitate

D. Rezania (✉) · T. Sasso
Gordon S. Lang School of Business and Economics, University of Guelph, Guelph, ON N1G 2W1, Canada
e-mail: drezania@uoguelph.ca

C. Machado and J. P. Davim (eds.), *Coaching for Managers and Engineers*, Management and Industrial Engineering,
https://doi.org/10.1007/978-3-030-71105-4_2

his or her becoming a more effective leader" (p. 4). For Harper (2012), leadership coaching prioritizes the growth of the coachee while also incorporating the ways in which a coach supports a coachee to develop leadership skills in alignment with the organization's mission. Approaches to coaching may strive to achieve different outcomes, from self-improvement and learning to improved work performance. Although people have differing views of what exactly coaching is and how it differs from other helping professions, such as counseling and mentoring, we note that, normally, coaching individuals (Hackman and Wageman 2005; Goleman et al. 2002):

- is a fairly short-term activity.
- consists of one-to-one developmental discussions.
- focuses on current and future performance/behavior.
- includes both organizational and individual goals, particularly learning and performance enhancement through action.
- works on the belief that clients are self-aware and do not require a clinical intervention.
- involves intentional change, a central element within many coaching methodologies and fundamental to the achievement of learning outcomes.
- is based on various assessments and use of a variety of tools and techniques from a wide range of theoretical backgrounds, including organizational theory, occupational psychology, psychometrics, learning, and counseling. Examples include learning style, and adaptive style inventories, 360-degree feedback, psychometric instruments, goal setting, and emotional intelligence models.
- is often based on the self-directed learning principle, in which individuals take the initiative to diagnose their learning needs, formulate learning goals, identify human and material resources for learning, choose and implement appropriate learning strategies, and control and monitor learning outcomes.

Although these items are characteristics of coaching individuals, we argue that they can be generalized to coaching teams, including those that would be responsible for project management (Rezania and Lingham 2009). To clarify this, we first consider the type of relationship between a coach and a project manage coachee, then we examine, and consider two current models of action research and process consultation. Finally, we consider some of the constructs used in coaching individuals and teams.

2 A Quadrangular Relationship Between the Organization, Coach, Project Manager, Project Team

The context of coaching leaders within an organization often involves a triangular relationship between the leader, the coach, and the organization. Organizations often define a set of desired managerial competencies and their associated behavioral manifestation. Examples include communication, persuasiveness, and sensitivity. The coach observes the coachee demonstrating those competencies, provides feedback on those behaviors, and creates action plans as a process for ongoing leadership development of the desired competencies of the coachee (Sherrer and Rezania 2020). In such instances, the triangular nature of the coaching relationship is clear right from the start. The coaching process focuses on the developmental objectives as defined by the organization, and the coach focuses on helping the project manager develop those competencies.

In contrast, coaching for personal development does not necessarily include an organization's objectives. In this coaching process, the relationship between the coach and the coachee is learning-focused, based on the power of positive thinking, and building on one's strengths (Goleman et al. 2002). The focus is more on the desire to change than to address behavioral problems. To develop competencies, a leader is supported through the coaching process to focus on their strengths, finding exceptions to the problems they identify as developmental areas and build on those (Boyatzis 2006).

Whether the focus is on personal development, development of competencies or behavioral alignment with organizational objectives, a central postulate of leadership development exists: a leadership coaching program has to systematically guide a client through integrating their knowledge, skills, and experiences to facilitate the desired change and learning. In this "wholistic" model, the leadership development program fosters the interrelationship between the content, the process, and the developmental change (Benson 1991). This emphasis centers focus on a person's behaviors, attitudes, and values in relation to work, people, and self.

Although we have utilized the triangular representation of the coaching relationship, it is possible to expand the focus. An important element of leadership development is to help leaders develop their followers (Bass 1990). In addition to a project manager learning to adjust their behaviors, attitudes, and values in relation to work, people, and self, project managers' performance is inherently dependent on the performance of their team (Keller 1986; El-Sabaa 2001; Anantatmula 2010). Therefore, project managers need to develop their team members to make more effective and better teams. Coaching project managers cannot only focus on their behaviors, attitudes, and values in relation to work, people, and self, but it must also consider helping them to work with their temporary teams (Rezania and Lingham 2009). Current research on factors contributing to team performance has focused largely on the functional and behavioral nature of teams, and emphasized themes such as learning (Edmondson 2003, 2002), commitment (Bishop et al. 2005) functions, affection and action (Hackman and Wageman 2005; Bryson and Bromiley 1993), communication (Keller 2001), interaction (Roberts et al. 2005), trust (Sheppard and Sherman 1998), vision (Dale C. and Derek H. T. W. 2004), and conversation (Baker et al. 2005). These constructs underscore the complexity of team effectiveness and performance management. Furthermore, projects are, per definition, temporary organizations and project leaders lead temporary teams, thus adding complexity around the extent to which follower development is within the capacity of project managers.

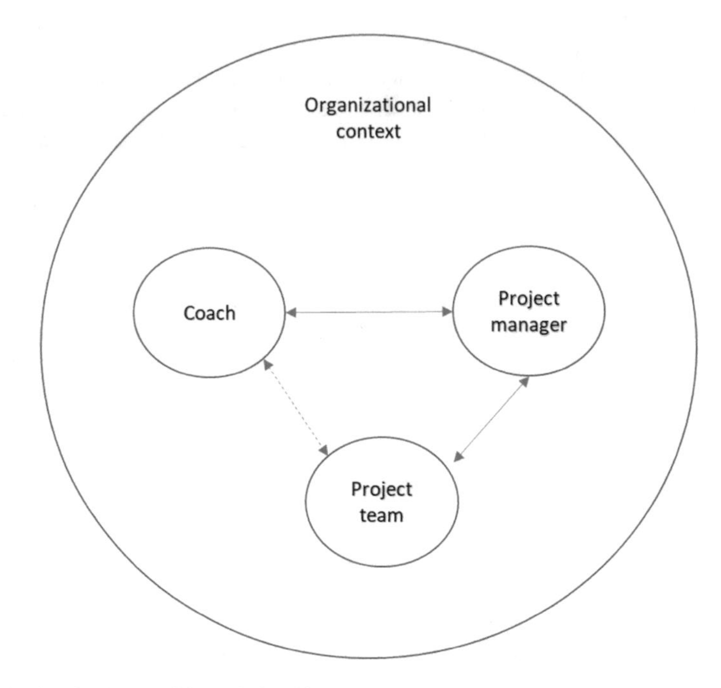

Fig. 1 Project leader coaching relationships

Despite the added complexity that comes with project members, as we conceptualize the building blocks of coaching project managers, we should include coaching focused on project managers and their project teams. This nuanced framework will more readily translate to the realities facing project managers and organizations by fully taking into account the team structure and dynamic.

Figure 1 presents the complex relationship between coach, project manager, project team and the organization. It is worth noting that the relationship between the coach and project manager should be fundamentally different than the relationship between the project manager and the team and the coach and the team. The coaching relationship is at the heart of the definition of coaching since it is through this relationship that the coach can challenge their clients 'comfort zone' and thereby support behavioral change and ultimately, transformation (Bluckert 2005; Passmore and Fillery-Travis 2011). At the individual level, the coaching relationship is a 'designed alliance' that is developed by both the coach and coachee (Vaartjes 2005). At the team level, team coaching is considered an act of leadership (Hackman and Wageman 2005). In the complexity of the dyadic relationship, the coach is helping the project manager to develop. Part of the development of the project manager is focused on learning to facilitate team development. The coach is indirectly working with the team through their relationship and dynamic with their client (the project manager).

2.1 Coach–Project Leader Relationship

There are various approaches to coaching individuals. Intentional Change Theory, which is based on self-directed learning, is one approach that could be employed in the coaching of project leaders. Knowles (1975) defines self-directed learning (SDL) as a process in which individuals take the initiative, with or without the help of others, in diagnosing their learning needs, formulating learning goals, identifying human and material resources for learning, choosing and implementing appropriate learning strategies, and evaluating learning outcomes. Intentional change is composed of five steps—also called discontinuities (Goleman et al. 2002). A discontinuity marks a moment of awareness, a sense of urgency. This step is considered essential for learning to proceed. Boyatzis (2006), Boyatzis and Akrivou (2006) identified five points of sudden realizations that are facilitated in a coaching process:

1. A person discovers their Ideal Self—personal vision, what they want out of life, or what kind of person they want to be;
2. A person discovers their Real Self—how they come across to others, which leads to a sense of strengths and weaknesses and how their actions compare to their Ideal Self;
3. A person discovers their agenda for the coming time period;
4. A person discovers ways to experiment and practice moving toward their Ideal Self;
5. A person discovers trusting, supportive relationships that help them through the process.

Through self-directed learning (going through the process of intentional change by realizing the leader's real and ideal selves) and co-creating an individualized learning plan with a coach, a leader is better situated to develop their ideal self. The process starts with a series of exercises to help the project leader identify their ideal self, where they want to be, and to catch their dreams and engage their passion, personally or professionally. Knowing one's ideal self marks the first discontinuity or moment of awareness. The second moment of awareness is understanding the real self. A multi-rater 360 feedback process, based on emotional intelligence (EI) competencies, can be part of this process. Project leaders learn that EI competencies are learned abilities, each of which has a unique contribution to making leaders more resonant and effective (Goleman et al. 2002). In addition, the coach works with the project leader to help them to understand and become aware of their fundamental values and philosophical orientations (Goleman et al. 2002). Through the first two discontinuities, the project leader articulates their strengths, the gap between their real and ideal selves, and their choices about what they want to change. The step focuses on the realization that everyone has a profile of strengths as well as areas to develop in the competencies.

The third discontinuity raises project managers' awareness about their need to develop a learning agenda. They focus on their desired future by looking at the gaps identified in the first two steps and setting personal performance goals. This plan requires an understanding of the emotional intelligence model: each competence uniquely contributes to performance, and certain competencies may interact and build on one another (Goleman et al. 2002).

The fourth discontinuity moves the project manager through the learning cycle, experimenting with the new behaviors and then practicing them. Here the project manager uses what they have learnt from assessing their learning style. Kolb (1984) describes learning styles as individual differences in learning based on the learner's preference for employing different phases of the learning cycle. In his theory of experimental learning, Kolb (1984) defines four learning styles: Converger, Diverger, Assimilator, and Accommodator. An individual with a convergent style prefers to grasp information through abstract concepts first, and then to transform this information through active experimentation. In contrast, divergers prefer to grasp information through a concrete experience first, and then to transform this experience through their own reflection. Awareness of learning style is important for project managers coaching process because if the learning agenda is not aligned with their preferred learning style (unless part of the plan is to develop your learning flexibility), they won't be comfortable with the learning process and will soon lose the motivation to continue.

In his experiential learning theory, Kolb states that integrative development is the ability to respond flexibly to various learning situations and demands, and has a strong influence on development and growth. Adaptive flexibility as a trait can be measured by the Adaptive Style Inventory (ASI) (Boyatzis 1993). People with high degrees of adaptive flexibility can readily adapt their learning styles to the demands of the situation. An adaptively flexible leader can more successfully manage in the global marketplace than a leader who is not. Through ASI assessment, a project leader can be coached to understand their reactions to the demands of a situation and to combine this understanding with their learning style to design effective practices.

The fifth discontinuity focuses on the need to create a supportive environment and relationships that enable project managers to learn. Such relationships create the context for change and help them to make sense of the learning process and appreciate their new identity. Project managers should have the opportunity to engage in learning practices supported by their coaches.

In this process, the relationship between the coach and the project leader is learning-focused, based on the power of positive thinking, and building on one's strengths (Goleman et al. 2002). The focus is more on the desire to change and improve performance aligned with project managers' objectives, which are impacted by the organizational objectives and project managers' dependence on their teams' performance. To develop competencies, the project leader is helped to focus on their strengths, find exceptions to the problems they identify as development areas and build on them. It is a purposeful and structured approach, helping them to realize that both real and ideal selves are socially constructed and that there is a developmental dimension to a person's capacity to reconstruct their world (Boyatzis and Akrivou 2006). Merely sharing information and engaging the project leader in activities in the traditional feedback setting will not develop these emotional intelligence competencies. Only through self-directed learning (going through the process of intentional change by realizing their real and ideal selves) and co-creating an individualized learning plan with a coach can they develop their ideal self.

2.2 Project Leader–Team Relationship

The relationship between the project leader and the team can be viewed as a mode of action research. Schein (1995) identifies two modes of action research. One method of action research is an extension of formal research methods in which the client system becomes involved both in the gathering and analysis of the data pertaining to the change problem. Through this method, the client becomes more of a collaborative researcher. Another mode of action research is an extension of clinical work where the client in the process of seeking help begins to engage in inquiry and research processes with the help of the consultant clinician, thereby making the clinician more of a researcher.

Schein argues that these two models of action research are fundamentally different in terms of their underlying assumptions. In the first model of action research, it is the researcher who wants to figure out how to be more successful in implementing some changes that they desire. By involving the targeted population in the research process, the population became more open to inquiry and committed to the desired change. The researcher involves the client in the researcher's issues. In this model, it is not the client's need that initiates the process, but the researcher's choice to involve the client. In contrast to this, in the "clinical" model, it is the needs of the client which drives the process and it is the client who involves the researcher in the client's issues. This model of "action research" is fundamentally different in that the initiative remains at all times with the client (Schein 1995).

This second model is fundamental for "process consultation" which advocates two principles (Schein 1988). First, the process of how people interact to get things done is critical to team effectiveness. Second, dysfunctional processes can be remedied through purposeful intervention. In this approach, the consultant (project leader) engages team members in analyzing group processes on two levels simultaneously: (1) the substantive level and (2) the internal level (Schein 1995). These two levels facilitate a better understanding of how interaction processes foster or impede effective group functioning. The intervention starts with the coach directly observing the group as it works on a substantive organizational problem. When the group is ready, the coach uses systematic or confrontive interventions intended to help the team exploit previously unrecognized opportunities (Schein 1995).

At the internal level, teams develop norms, values, and beliefs for their group-related behavior (Edmondson 1999; Katzenbach and Smith 1993). Working in teams is experienced based on these group-level norms, values, and beliefs. Therefore, the coaching process that the project manager is going to design must also deal with the interactions that team members experience (Rezania and Lingham 2009a; b). The first step of the coaching process assumes that interaction is a necessary condition for team learning to happen and that learning is central to any intervention (Lingham 2004).

Given the importance of team interaction, the coach can help the project manager to pay attention to what kind of interaction the team wishes to have. Usually, the actual experience of interaction is not the ideal that team members had hoped for. There is a discrepancy between what team members experience and what they wish to experience. The self-discrepancy theory says that we are strongly motivated to maintain a sense of consistency among our various beliefs and self-perceptions (Higgins 1999). The difference between our aspirations for ourselves and our actual behaviors results in various feelings such as sadness, dissatisfaction, and other anxieties (Higgins 1989). In these situations, we act to reduce this discrepancy by various means.

The second component of the coaching process uses this discrepancy between a team's real and ideal experiences to trigger self-reflection (Rezania and Lingham 2009). Teams need to identify where they are and where they would like to be (such as their stage of team development, existing behaviors, ideal states, or level of performance) (Rezania 2008).

Since interaction is a necessary condition for team learning and development, coaches can help project managers to juxtapose information about a team's actual (real) interaction with their ideal interaction. There are various methods for doing this. Teams need to know where they are (whether it be stage of team development, level or height of performance, or existing behaviors) and where they would like to be (ideal states or level of performance). This assumption has been foundational for various approaches to coaching individuals (Boyatzis and Akrivou 2006). One instrument, the Team Learning and Development Inventory (TLI), measures actual and ideal experiences of teams across four dimensions of teamwork (Lingham 2004). By using an instrument like the TLI, the project manager can help their team to create an awareness of their real and ideal interaction. The awareness enables a team to engage in team-directed learning and to identify its ideal space. The project leader can then help the team to create their ideal space (or important aspects of it). This method encourages team leaders to provide a context in which the right teamwork space is created.

The third component of the coaching process is similar to Boyatzis' self-directed learning method (Boyatzis 2006). An effective methodology to facilitate team-directed learning and development would

(1) require teams to engage in developing team-level awareness;
(2) provide teams with a common language around team interaction;
(3) provide feedback on the quality of current team interaction;
(4) allow team members to craft concrete action steps and practice new behaviors, to improve the quality of their team interaction and performance.

This third component requires the project manager to trigger a self-directed learning process for the team. Intentional change at the team level is central to the achievement of learning outcomes, a process in which teams take the initiative in diagnosing their learning needs, formulating learning goals, identifying human and material resources for learning, choosing and implementing appropriate learning strategies, and controlling and monitoring learning outcomes.

3 Principles for Designing the Facilitation Process of Coaching Project Leaders and the Team

Based on the preceding review of the literature and an understanding of the complex dynamic of coaching as it applies to project managers and their ability to successfully lead project teams, we have identified four core principles that can shape project manager coaching. These principles provide actionable and meaningful direction for supporting the leadership context of project managers through an integration of dynamic that is currently under-researched and supported in practice.

Principle One: Support collaborative construction of knowledge through social negotiation between the coach, project manager, and project team, and the organization. People are engaged in an ongoing process of making sense of the world through interaction with others (Schutz and Collected papers 1962). A key concept in understanding how people make meaning is intersubjectivity. In other words, we base our thoughts about the world on our experience of the world, and our experience is intersubjective because we experience the world with and through others (Weick 1995). A coach is not a detached observer, but they are someone who supports the collaborative construction of knowledge through social negotiation.

Principle Two: The focus of the coaching session is not on giving the project manager or the team information about what constitutes being a good project manager or a good team but on knowledge construction, exploration of actual experiences, and construction of a path towards an ideal project manager and ideal team. The process to initiate or sustain change starts with exploring what is an ideal project manager and an ideal team in the context of the organization. Why? Individuals and teams have developed norms and habits which are not easy to change. Engaging the project manager and the team in exploring its ideal self will energize them and give them the power to move forward (Boyatzis and Akrivou 2006). The objective is not to tell teams how they should behave, but to help them negotiate what is their ideal and how to behave based on their actual experiences.

Principle Three: The process should explore the actual experiences of the project manager and the team. It is based on the context in which they function rather than abstract experiences or role play. Roleplay has traditionally been used as a very flexible and effective tool in training individuals and teams. It helps individuals to understand the theory and to transfer concepts into practical experience. However, the objective of coaching is not to give them abstract concepts of effective project management or teamwork, instead of the coaching process should help them to explore those concepts. Only after a project manager or a team has decided what its ideal is, can roleplay and similar tools be useful to help them practice new behaviors.

Principle Four: The process cultivates reflective practice. Similar to experiential learning for individuals, Gibson and Vermeulen (2003) conceptualize team learning as "a cycle of experimentation, reflective communication, and knowledge codification" (p. 222). In the process that we use, all three processes should be present. Learning is then the product of these three factors: experiencing things, evaluating together, and naming what one has learned.

4 Constructs Used in Coaching Project Leaders

The content of a project manager's unique coaching experience can involve addressing many individual-level competencies and skills. However, as previously noted, the coaching of project managers may involve greater complexity with the role of team experiences and the desire to engage in temporary follower development. As such, coaches and project managers should be conscious of common individual-level coaching constructs and the extent to which these operate at a group-level. Transferring some of these constructs, such as insight, self-awareness, goal setting, potency, problem-solving, self-evaluation, cognitive change, behavioral change, and so on from the individual to the team level may seem straightforward at first. However, unlike the relatively simple relationships at the individual level, group-level relationships, constructs, and effectiveness are complex and seem to be moderated by other multi-level and multi-source factors in the workplace. Individual-level constructs do not depend on the processes of interacting, exchanging, and combining information, or on integrating and processing collective cognition.

5 Conclusion

Coaching project managers can be critical for leading projects to success. Why? Part of a project team leader's function is to help the team develop to their ideal context, based on their own experiences and needs. When coaching project managers includes helping them to coach their teams, they enable teams to create an awareness of their real and ideal experiences by measuring and mapping both of these spaces, and to engage in team-directed learning. Coaching can also assist the project leader in enabling the team to create this ideal space (or important aspects of it) as identified by them. We know that teams who are aware of a gap between their real and ideal spaces act to fill this discrepancy. Using the method of developing towards a team's ideal, we can coach team leaders to provide a context in which the right conversational space is created.

Using this method of coaching helps a team to convert its experiences into words and categories. This process constitutes learning by turning unspoken knowledge into text, which forms the basis for further discussion and action. In the cases presented, we observe teams detecting what can be improved. According to organization theory, detecting and correcting errors is an active process of organizing.

When a project manager is coached based on this framework, they help their team to develop its self-regulatory processes and learn how to learn. In essence, this intervention helps the team to improve its metacognitive skills (learning how to learn), which in turn regulates its interaction towards performance. Most researchers believe that both the individual and shared mental models develop and are maintained by gathering information during the phase in which teams monitor their actions and regulate them (Lord 2000). Therefore, this approach to coaching project leaders may enhance project leaders' and their team performance directly as well as indirectly, through its impact on the development of shared mental models, shared norms, and shared beliefs.

References

Anantatmula, V. S. (2010). Project manager leadership role in improving project performance. *Engineering Management Journal, 22*(1), 13–22.

Baker, A. C., Jensen, P. J., & Kolb, D. A. (2005). Conversation as experiential learning. *Management Learning, 36*(4), 411–427.

Bass, B. M. (1990). *Handbook of leadership.* New York: Free Press.

Benson, L. (1991). *A youth leadership training case study.* University of Alberta.

Bishop, J. W., et al. (2005). A construct validity study of commitment and perceived support variables: A multifoci approach across different team environments. *Group and Organization Management, 30*(2), 153–180.

Bluckert, P. (2005). *Critical factors in executive coaching–the coaching relationship.* Industrial and Commercial Training.

Boyatzis, R. E. (2006). An overview of intentional change from a complexity perspective. *Journal of Management Development, 25*(7), 607–623.

Boyatzis, R. E., & Akrivou, K. (2006). The ideal self as the driver of intentional change. *Journal of Management Development, 25*(7), 624–642.

Boyatzis, R.E., & Kolb, D.A. (1993). Adaptive style inventory: Self scored inventory and interpretation booklet (Journal Article).

Boyatzis, R. E., Smith, M. L., & Blaize, N. (2006). Developing sustainable leaders through coaching and compassion. *Academy of Management Learning & Education, 5*(1), 8–24.

Bryson, J. M., & Bromiley, P. (1993). Critical factors affecting the planning and implementation of major projects. *Strategic Management Journal, 14*(5), 319–337.

Delantey, J. T., & Huselid, M. A. (1996). The impact of human resource management practices on perceptions of organizational performance. *Academy of Management Journal, 39*(4), 949.

Edmondson, A. (1999). Psychological safety and learning behavior in work teams. *Administrative Science Quarterly, 44*(2), 350–383.

Edmondson, A. C. (2002). The local and variegated nature of learning in organizations: A group-level perspective. *Organization Science, 13*(2), 128–146.

Edmondson, A. C. (2003). Framing for learning: Lessons in successful technology implementation. *California Management Review, 45*(2), 34–54.

El-Sabaa, S. (2001). The skills and career path of an effective project manager. *International Journal of Project Management, 19*(1), 1–7.

Ely, K., et al. (2010). Evaluating leadership coaching: A review and integrated framework. *The Leadership Quarterly, 21*(4), 585–599.

Gibson, C., & Vermeulen, F. (2003). A healthy divide: Subgroups as a stimulus for team learning behavior. *Administrative Science Quarterly, 48*(2), 202–239.
Goleman, D., Boyatzis, R. E., & McKee, A. (2002) *Primal leadership: realizing the power of emotional intelligence.* Boston: Harvard Business School Press, p. 306.
Guest, D. E. (1997). Human resource management and performance: A review and research agenda. *International Journal of Human Resource Management, 8*(3), 263–276.
Hackman, J. R., & Wageman, R. (2005). A theory of team coaching. *Academy of Management Review, 30*(2), 269–287.
Harper, S. (2012). The leader coach: A model of multi-style leadership. *Journal of practical consulting, 4*(1), 22–31.
Higgins, E. T. (1989). Self-discrepancy theory: What patterns of self-beliefs cause people to suffer? In L. Berkowitz (Ed.), *Advances in experimental social psychology* (pp. 93–136). New York: Academic.
Higgins, E. T. (1999). When Do self-discrepancies have specific relations to emotions? the second-generation question. *Journal of Personality and Social Psychology, 77*(6), 1313–1317.
Katzenbach, J. R., & Smith, D. K. (1993). *The wisdom of teams: Creating the high-performance organization.* Boston Mass: Harvard Business School Press, p. 291.
Keller, R. T. (1986). Predictors of the performance of project groups in R & D organizations. *Academy of Management Journal, 29*(4), 715–726.
Keller, R. T. (2001). Cross-functional project groups in research and new product development: Diversity, communications, job stress, and outcomes. *Academy of Management Journal, 44*(3), 547–555.
Knowles, M.S. (1975). *Self-directed learning.* New York: Association Press, p. 135.
Kolb, D. A. (1984). *Experiential learning: Experience as the source of learning and development.* Englewood Cliffs, NJ: Prentice-Hall, p. 256.
Lingham, T. (2004). Developing a measure for conversational spaces in teams. Unpublished doctoral dissertation. Cleveland, OH: Case Western Reserve University.
Lord, R. G. (2000). Thinking outside the box by looking inside the box: Extending the cognitive revolution in leadership research. *Leadership Quarterly, 11*(4), 551.
Passmore, J., & Fillery-Travis, A. (2011). A critical review of executive coaching research: A decade of progress and what's to come. *Coaching: An International Journal of Theory, Research and Practice, 4*(2), 70–88.
Rezania, D. (2008). A framework for team coaching: Using self discrepancy theory. *Development and Learning in Organizations: An International Journal.*
Rezania, D., & Lingham, T. (2009a). Coaching IT project teams: A design toolkit. *International Journal of Managing Projects in Business, 2*(4), 577–590.
Rezania, D., & Lingham, T. (2009b). Towards a method to disseminate knowledge from the post project review. *Knowledge Management Research & Practice, 7*(2), 172–177.
Roberts, T. L., et al. (2005). The effects of information technology project complexity on group interaction. *Journal of Management Information Systems, 21*(3), 223–247.
Schein, E. H. (1988). *Process consultation.* Addison-Wesley series on organization development, vol. 2. Reading Mass: Addison-Wesley.
Schein, E. H. (1995). Process consultation, action research and clinical inquiry: Are they the same? *Journal of Managerial Psychology, 10*(6), 14–19.
Schutz, A. (1962). *Collected papers.* Phaenomenologica, vol. 11, 15, 22. The Hague: M. Nijhoff.
Sheppard, B. H., & Sherman, D. M. (1998). The grammars of trust: A model and general implications. *Academy of Management Review, 23*(3), 422–437.
Sherrer, M., & Rezania, D. (2020). A scoping review on the use and effectiveness of leadership coaching in succession planning. *Coaching: An International Journal of Theory, Research and Practice*, 1–11.
Thomas, J., & Mengel, T. (2008). Preparing project managers to deal with complexity–Advanced project management education. *International Journal of Project Management, 26*(3), 304–315.

Vaartjes, V. (2005). Integrating action learning practices into executive coaching to enhance business results. *International Journal of Evidence Based Coaching and Mentoring, 3*(1), 1–17.
Weick, K. E. (1995). *Sensemaking in organizations.* Foundations for organizational science. Thousand Oaks: Sage Publications, p. 231.

Socialization Coaching: An Application for Welcoming and Integrating People ("The Point Out of the Curve") Without Homogenizing

Moacir Rauber and Carolina Feliciana Machado

Abstract It is proposed to use of coaching for the encounter phase of the socialization of new employees as another resource for managers and the human resources area. It is a socialization tactic predominantly individualized and informal in context; non-sequential and variable with respect to content; disjunctive, by individualized classification, and investiture, regarding institutionalized classification, in relation to the social aspect. Socialization coaching is an application to treat people as complete, complex and interdependent systems, understanding each individual as a "Point Out of the Curve" since their arrival at the organization. Coaching, as a tool for the socialization of employees in the organization, expands the possibilities of welcoming by encouraging the active movement to seek to integrate. Recognize the other as a point outside the curve without discarding it, as sometimes indicated in statistics; and integrate it into the organizational environment without homogenizing it, as often happens in the socialization proposed in the learning trails; is the challenge of organizations that seek to maintain competitiveness.

1 Introduction

Here we propose the use of coaching to the encounter phase of the socialization process of new employees, contributing with the development of one more resource for managers and the Human Resources Management (HRM) area. A resource to lead with people as a complete, complex and interdependent system, understanding each individual as a point outside the curve since their arrival at the organization, a delicate phase in the life of the organizational employee. For this reason, coaching being offered at the time the employee enters the new organization aims to expand

M. Rauber
School of Economics and Management, University of Management, Braga, Portugal

C. F. Machado (✉)
School of Economics and Management, University of Minho, Interdisciplinary Centre of Social Sciences (CICS.NOVA.UMinho), Braga, Portugal
e-mail: carolina@eeg.uminho.pt

C. Machado and J. P. Davim (eds.), *Coaching for Managers and Engineers*, Management and Industrial Engineering,
https://doi.org/10.1007/978-3-030-71105-4_3

the possibilities of welcoming and integration by encouraging the active movement to seek integration. Recognize the other as a point outside the curve without discarding it, as sometimes indicated in statistics; and integrating it into the organizational environment without homogenizing it, as it often happens in socializing in the welcoming and integration proposals present in the learning trails, is the challenge of the organizations that seek to maintain competitiveness.

In statistics, usually, the point outside the curve is treated as a discrepancy, anomaly, abnormality, exception, or failure. In quantitative data, the points outside the curve can be a typo or a poorly collected sample, making it relevant to discard them even before calculating the averages, if the objective is to set parameters. However, it is surprising to see decisions about people being made using statistical techniques, including when discarding points outside the curve. The data, or rather, the people are grouped into graphs and spreadsheets that indicate the statistics that are used to make a decision. A chart with responses from more than 100 candidates has two or three points outside the curve that do not align with the responses expected for a given situation or profile. Without knowing who the points outside the curve are, they are discarded, without understanding that each person is a complete, complex, and interdependent system.

In the defended point of view, using statistical information about people to then classify them and create a taxonomy can be a disservice to managers and to people management. Grouped data for a given group of people can indicate a trend, but it cannot classify individuals. It is understood that all are plural, unique and singular individuals, being naturally a point outside the curve. There is no way to be equal by being different. We do not see the possibility that a given response could lead to the person being classified as resistant to change, reactionary, not very creative or without leadership skills. The person can be all this until the day he ceases to be. Neuroscience confirms that the human being has the ability to develop, improve, and modify behavior at any stage of his life. Looking back, and also nowadays, many managers insist on classifying and labeling people considering their date of birth or their social origin. It is understood that there may be a behavioral tendency among those born in a given period due to their age and the coexistence they had, but it is not agreed to classify people in this way, because injustices are simply committed. And injustices affect people, who are not mere data, and organizations, which only exist with people. How to say that someone 50 years old is less creative than someone who is 20 years old simply because of their date of birth? Why validate the idea that a person under thirty seeks less stability than someone over forty?

Using statistical information about people and their behavior to validate competencies is a risk. There are countless the factors that can influence an answer that indicates that someone has a leadership profile or a highly creative vision. People are not the same and they are not static. They are always on the move even though they are apparently not moving. It is not possible to classify human beings, because there is no statistic that covers all the variables that a human being is subject to in a day, imagine throughout an organizational life that begins with the arrival of the employee. It is understood that despite the apparent human rationality, the human being is not always logical, transforming him into a point outside the curve. So the question: what to do with the point outside the curve? It is believed that it is important for each manager to remember that he is not a point, but a person. This can make the difference, positively. Offering coaching as an alternative to socialization known as the meeting phase, tends to contribute to a welcoming and integration process in which the individual plays an active role when being recognized and accepted as a point outside the curve for its uniqueness that is welcome.

2 The Point Out of the Curve: The Reasons for Coaching That Acknowledges And Accepts People Active in the Socialization Process

According to Allen (2006), new employees leave companies frequently before they even feel like members of the organization, because they have not completely adapted to organizational values and culture. Mosquera (2007) also states that the turnover in organizations is significantly greater in the period of entry of the individual in the organization than in later periods. It is the phase of meeting the new employee with the new organization, a period coinciding with the one in which socialization practices must be thought out for the reception and integration of employees (Griffeth and Hom 2001). In it, there is often a breach of psychological contract, a high rate of unsuitability for new employees and, consequently, layoffs due to requests and also low productivity. The same statement is shared by Allen (2006) who believes that it is a problematic issue for organizations, because there were made investments in recruitment, selection, and training, causing losses by not recovering the investment made. According to Allen (2006), one of the main reasons for the early departure of new employees is the inadequate socialization process.

For Kim et al. (2005) socialization is fundamental for the adjustment of the person to the organization, since the first objective is to provide a picture of the culture and organizational values that enable new employees to respond positively to the environment, together with other employees. Previous works, such as those carried out by Van Maanen and Schein (1979), warned to the same problem. It must be understood, according to the authors that new employees when joining the organization must deal with their expectations and the real environment they found. Likewise, new employees do not have yet a comfortable routine that allows them to interact with others, predicting possible responses (Van Maanen and Schein 1979). Finally, however adjusted and clarified the recruitment, selection, and job description process may have been, there is a lack of identification with the job, as well as with what happens around them (Gruman et al. 2006). Thus, according to Cable and Parsons (2001), how organizations treat new employees in the first months sends clear signs of how well they can adjust to it or not. Each individual is unique and can be considered as a complete, complex, and interdependent system, just as it is a point outside the curve. How he will be treated in the reception and integration will make a difference on his way to become a full member of the organization.

It is understood, therefore, that an adequate socialization process in the encounter phase tends to decrease turnover, improving the level of adjustment of new employees to the organization, as supported by Cable and Parsons (2001). Kim et al. (2005) also state that the implementation of well-structured socialization strategies tends to reduce ambiguity and anxiety for new employees, but it can stimulate passivity by assuming that the organization is in charge of conducting the process. The understanding that the organization controls the entry process has been challenged by the proactiveness of new employees who, according to Griffin et al. (2000), can play an important role in the socialization encounter phase. Proactivity is one of the main skills expected from new employees by employers, according to research conducted by Huhman (2014). It is cited as proactivity, in addition to having its characteristics found in denominations such as self-motivation, initiative, adaptability, creativity, innovation, and the ability to solve problems. Proactivity tends to lead new employees to seek information about the environment, tasks, and organizational routines with the new network of contacts, with former employees, in welcoming manuals and other sources that allow them access to rules and the organizational culture (Huhman 2014).

Stimulating the proactivity of new employees, considering all the skills previously mentioned, is part of the justification for the introduction of coaching in the encounter phase of the socialization process as an individualized integration tactic when recognizing the new employee as a “real other”. A point outside the curve. Understanding that coaching is a way of intervention that allows the individual to assume responsibility for the development and improvement of personal performance, contributes to understand that we have a tool that encourages individual proactivity with positive organizational results (Gormley and van Nieuwerburgh (2014). Coaching is considered a fast and effective way of human development with the results focusing on the one who participates in the process, improving performance, reducing turnover, and increasing individual productivity, with reflexes in organizations (Neale et al. 2009).

Similar concepts are presented by authors who are considered coaches of excellence in the global market. Wolk (2010) states that coaching is an interaction process that allows the individual to play an active role in the search and achievement of his goals, allowing him to develop their full potential. Whitmore (2009) and Downey (2010) also describe coaching as the job of facilitating learning, stimulating development, and increasing the performance of individuals by unlocking their potential, with positive effects on the organizations to which they belong. This means that it is up to the manager to include the point outside the curve and not simply discard as in the statistic that works with data.

When proposing the use and application of coaching in the encounter phase of the socialization process, it tends to encourage the proactivity of new employees who become active elements in the process, according to Griffin et al. (2000), and to decrease the effects of leveling, very common in institutionalized and structured processes of socialization (Herrmann 2013). This author, who studied Kierkegaard and indirect communication, understands leveling as a discursive and communicative process that reduces the individual to an element of mass, very common in institutionalized socialization processes.

With the use of coaching since the arrival of the new employee to the organization, the process of self-learning and guided self-learning is also stimulated, since the coaching will be used for specific purposes of interest to the new employee (Ellinger 2004). Therefore, the introduction of coaching by management in the socialization encounter phase is intended to encourage the new employee to assume part of the responsibility for his socialization process, becoming an active element of the process and not just its object. Part of the logic of socialization is inverted, which in the current view starts from the assumption that the new employee must be integrated, leading him to play a role of wanting to integrate, proposed by the application and use of coaching in the socialization process.

In this way, the vast majority of tactics and tools used so far in the socialization encounter phase understand that the new employee is socialized, which induces him to be a passive element in the process. The proposition of applying and using coaching in the socialization encounter phase makes the new employee an active part in the process, stimulating and developing his proactivity in a complementary process that can make socialization faster and more effective. In this way, it is intended to anticipate the results obtained by using coaching as a personal development tool for employees found in other phases of their organizational life.

It is accepting the point outside the curve represented by each individual in the organization, who understands that are the specificities and singular and unique characteristics of its employees that create a competitive differential. The multiplicity of diversity is what makes the organization unique, a complete, complex and interdependent system formed by unique individuals who are complete, complex and interdependent systems. Offering coaching as an alternative for welcoming and integration in the socialization encounter phase is to exhibit a systemic view as manager of the organization and the individual.

Research cited by Neale et al. (2009) show that 92% of the organizations interviewed in the United Kingdom use coaching with their employees and the results show the rate of 96% of respondents who consider that there has been an improvement in individual performance, with a consequent increase in organizational performance. Thus, the surveys above confirm the effectiveness of the use of coaching by organizations as a personal development tool. However, there are no references regarding the use of coaching in the socialization process of new employees as a welcoming and integration tactic.

It is worth mentioning, therefore, that coaching is used throughout the career of employees, especially when it is perceived a potential to be developed or a conflict to be managed. One can even cite an example from Barner and Higgins (2007) that confirms the importance of offering coaching as a socialization tool in the phase of encounter of new employees. Researchers describe the situation of the hiring of a new sales director by a pharmaceutical industry that, six months later, was referred to a coaching process. This is because despite the excellent credentials and the right sales strategies proposed by the new director, he had a series of conflicts in his new activity. Due to these difficulties, it was suggested to the new director to work with an external coach to minimize conflicts. Therefore, had the new director entered the organization with the possibility of integrating himself through the use of coaching, perhaps he would not have generated the conflicts he faced, almost motivating his departure from the organization.

In the literature review done, there were no references to coaching being intentionally used by the HRM as a tool for socializing new employees, regardless of their hierarchical level. Likewise, interviews with HR managers showed its absence. Thus, there was a gap in the use of coaching as a tool for personal development within organizations. For this reason, a new application of coaching is advocated as a personal development tool aimed at the moment of the arrival of new employees in the organization as another welcoming and integration tactic. It is a proposal to treat each individual as a unique human being, a point outside the curve that will contribute with its uniqueness to the organization. Therefore, the socialization coaching proposal is inserted in the methodologies, it feeds on the theories present in the different coaching approaches and models. The intention is that managers and the HR area of organizations use coaching since the encounter phase in the socialization of new employees, as an individualized strategy and complementary to institutionalized tactics. With this, it is possible to anticipate the development of potential, prevent conflicts, and improve individual and organizational performance, reducing the need to "discard the points outside the curve" when understanding that they can be the competitive advantage of the organization.

It is understood that the use of coaching can allow the development of a new way of thinking about the socialization process of individuals in organizations, taking advantage of the fact that HRM and socialization are centered on the organization, while the coaching process is centered on the individual. This also justifies the use of coaching as a personal development tool applied when new employees arrive at the organization as another welcoming and integration tactic, contributing to an adequate socialization process. It focuses on a complementary approach to socialization that favors organization, giving the individual an active role in the process. The intention is to reduce the passivity of the new employee as an element to be socialized, becoming an active element in the process when seeking socialization (Herrmann 2013). Coaching is also considered a tool that develops through self-learning, assuming the contours of self-direction if applied for a specific purpose of those who use it (Ellinger 2004).

It is intended that the employee is not only welcomed, integrated and socialized, but that he is integrated and socialized, playing an active role in his process of becoming a full organizational member, recognizing that he is a point outside the curve.

A new application of coaching is proposed as a personal development tool directed to the moment of the arrival of new employees in the organization as another welcoming and integration tactic. Receiving the other as a real other and recognizing that he is a point outside the curve has the objective of creating competitive advantage for the organization. So, we start from the following questions:

- *How to take advantage of the benefits of the coaching process as another welcoming and integration tool in the encounter phase in the socialization of new employees?*

- *What to do to welcome and integrate without homogenizing to take advantage of the singularities of each new employee?*

From the questions, the focus can be directed to enjoy the benefits of the coaching process. With this, the manager and the HR area can receive and recognize a new employee as a point outside the curve, understanding him as a complete, complex and interdependent system, increasing organizational competitiveness by giving the individual an active role in the process.

Thus, the objective of this chapter consists in propose the application of coaching in the phase of encounter in the socialization of new employees, as another individualized welcoming and integration tool, without prejudice to other practices and tactics that are designed for the process. Seeing, receiving, and recognizing the other as a real other, recognizing him as a point outside the curve, can give the organization and the manager a competitive advantage by taking advantage of the singularities present in the multiplicity of individuals, complete, complex, and interdependent systems.

3 Convergences and Complementarities of Coaching and Socialization That Accompanies the Active Choices of People Who are Integrated

Before presenting the application model with a step-by-step approach, the convergences and complementarities between the two main topics of the proposal are rescued: socialization and coaching.

On the one hand, it is worth mentioning that socialization is the path taken by an external element to the organization to become an internal element, feeling properly socialized after a process of knowledge, acceptance, and integration in the new organization (Cooper-thomas and Anderson 2006). Socialization can also be conceptualized as being the process and the way people who are already in an organization structure and organize the recruitment, the entry, and the movement of employees in order to allow them to become effective organizational actors, taking advantage of their experiences and knowledge (Van Maanen 1978; Van Maanen and Schein 1979). These concepts involve the phases of anticipatory socialization, socialization in the encounter phase and socialization in the changes and acquisitions phase. The present chapter focuses on the encounter phase.

On the other hand, coaching is understood as a personal development tool that takes shape in a process that uses focused and structured interaction strategies between coach and coachee, with techniques that intend to stimulate the changes that those who participate in the process want (Cox et al. 2010). Coaching is also considered a process that allows individuals to assume the role of their choices when making knowledge work for themselves on the path to developing their full potential (Wolk 2010). Anyway, coaching can be conceptualized as a way to stimulate learning and personal development, making people improve their organizational performance by unlocking their potential (Downey 2010; Whitmore 2009).

When analyzing the concepts of socialization and coaching, one can see the complementarity between what the first needs and what the second offers. However, until now, coaching has not been used in the socialization phase, which is the moment when the new employee arrives at the organization, as part of the welcoming and integration tools. Searching in the literature about the areas in which a socialization process should focus and the benefits that coaching can offer, complementarity is evident. On the one hand, a socialization process must comprise six areas, namely, (1) socialization tactics; (2) socialization training; (3) proactive socialization; (4) learning and socialization content; (5) group socialization; and (6) moderators, mediators and individual differences (Saks and Ashforth 1997).

On the other hand, the use of coaching in organizations offers the following advantages: (1) it contributes to the construction of people development programs in organizations; (2) operates in the performance management and organizational feedback processes; (3) offers a holistic approach to unlocking individual and organizational potential; (4) increases performance within organizations; (5) demonstrates an organizational commitment to individual growth; and (6) creates a favorable environment for changes, rewarding employees, customers, and other stakeholders (Gormley and van Nieuwerburgh 2014).

Figure 1 shows that there exists complementarity between what socialization understands in its areas of concern with the process and what coaching as a personal development tool has as intention to offer.

The six areas of coverage that must be present for a socialization process to be effective and align with the advantages of implementing coaching to contribute to this process (Saks and Ashforth 1997).

CONVERGENCES BETWEEN SOCIALIZATION AND COACHING	
SOCIALIZATION – Needs	COACHING – Benefits
Socialization tatics	People Development Program
Socialization Training	Performance and Feedback Management Process
Proactive Socialization	Holistic view and unlocking the potential
Learning and Socialization Content	Organizational Development
Group Socialization	Organizational commitment and individual growth
Moderators, Mediators and diferences	Favorable to changes

Source: Adapted from (Gormley & van Nieuwerburgh, 2014; Saks & Ashforth, 1997)

Fig. 1 Convergences between socialization and coaching

The socialization tactics, classified, according to Jones (1986), as being institutionalized and individualized, divided between the areas of context, content, and social aspects, have the function of integrating the new employee through the reduction of uncertainties in the new environment, as well as to stimulate the development of its creative, innovative, and proactive potential (Saks and Ashforth 1997). It can be seen that it is something inserted in what coaching proposes as a people development program, which can gain relevance in the socialization phase, since it is a period in which the new employee tends to confirm or not the established psychological contract. (Gormley and van Nieuwerburgh 2014).

When talking about socialization training, an intentional and organized process by the organization to integrate the new employee is in mind (Saks and Ashforth 1997). A planned, programmed, and organized process to also meet the demands of the employee, with positive effects on the organization by reducing the turnover index that generates costs and losses. Trainings work with the reduction of uncertainty and with the stimulation of competences, skills, and positive attitudes of the new employee. Coaching, on the other hand, offers the advantage of performing performance management, coupled with feedback on the situation in which the new employee finds himself, leading him to assume the role of his organizational trajectory (Gormley and van Nieuwerburgh 2014).

The third line of Fig. 1 above, which deals with proactive socialization, is a relevant item, since it was found that most socialization tactics focus on the process of formatting the new employee according to what is expected from him. However, socialization itself provides that the proactivity of the new employee must be sought as a way of ensuring that the unique characteristics of the new employee can be used by the organization (Saks and Ashforth 1997). And proactivity, creativity and innovation are much more linked to the desire to learn from the new employee, who is central to a coaching process. Coaching enables the new employee to develop a systemic view of the organization and of himself, leading him to unlock his potential, benefiting and benefiting the organization with the process (Gormley and van Nieuwerburgh 2014).

Figure 1, in line three, shows the contents and learning in the first column as an area covered by socialization. It is understood that the socialization through the contents and the learning process of the new employee must focus, primarily, on performance proficiency, politics, language, people, organizational values and objectives, and the history of the organization as elements of learning for the new collaborators (Saks and Ashforth 1997). If, on the one hand, institutionalized tactics tend to achieve better results with areas of performance proficiency, on the other hand, individualized tactics can more positively influence the adoption of values, organizational objectives, language and internal policy itself. Once again, the benefits produced by coaching give the vision of complementarity in the socialization process, providing the new employee with the possibility to demonstrate and guarantee organizational performance (Gormley and van Nieuwerburgh 2014).

Group socialization, the fifth item in Fig. 1, reveals a critical period within the encounter phase of the socialization of new employees, because before becoming an organizational member it is essential that he is accepted, and becomes a member of the work group. It is necessary that the new employee goes through joining the new group, being able from there, to engage, commit, and move forward to assume the effective role of member of the group, or not (Saks and Ashforth 1997). In this way, the new employee needs to win a game of support forces and resistance to his entry into the group. Some will try to modify him so that he is an element that contributes to the maximum, while the new employee will want to adjust the group to their expectations. Along this path, coaching, as an individual tool, can help the new employee to assume the organizational commitment aligned with the need for individual growth (Gormley and van Nieuwerburgh 2014). It is a complementary tool and can be a safety zone for the new employee to expose his fears, uncertainties, and insecurities that he faces in the group.

The sixth line of Fig. 1, which in the column of socialization talks about moderators, mediators, and differences, is a concern with the dissimilarities that can contribute or hinder an adequate socialization process. At that moment, the concern is with what makes the employee different, preventing him from being a full member of the new organization (Saks and Ashforth 1997). There are cultural variables, age, and lifestyle differences that most strongly impact adequate socialization and coaching, as a tool that benefits from change, tends to lead new employees not to judge the difference, but to understand it.

Therefore, it is clear that the six areas that are understood in an adequate socialization process are complementary in what coaching as a personal development tool has to offer, finding a juxtaposition shown in Fig. 1. It is that management offers the possibility that the new employee is received and recognized as a point outside the curve to also integrate, recognizing in others the same characteristics.

By punctually advancing towards socialization in the encounter phase, which involves welcoming and integration activities, Feldman (1976) establishes four processes that comprise it:

(1) initiation to the task, which occurs as the employee gains skills to fully perform his work;
(2) initiation to the group, which occurs through the acceptance of the new employee by the other members of the organization, indicating the evolution in the construction of their new interpersonal relationships;
(3) role definition, which occurs through an implicit or explicit agreement with the working group on what is the mission to perform, the priorities, and the allocation of time for these tasks, indicating the degree of clarification of its role; and
(4) congruent assessment, carried out by the employee and a supervisor assessing their progress in the organization, seeking consensus on the fact globally and also assessing specific strengths and weaknesses.

Comparatively, the transformational coaching methodology has four stages that converge with the four processes proposed for the socialization encounter phase. A transformational coaching process, as proposed by Wolk, (2010) has:

(1) introduction or opening, marks the beginning of the relationship between coach and coachee and then the beginning of the session and the process;

Socialization Processes in the Encounter Phase and Coaching Stages	
Socialization Processes – Encounter Phase	**Coaching Stages**
Task initiation	Introduction / opening
Group initiation	Exploration, understanding and interpretation
Role definition	Expansion
Congruent assessment	Closure

Source: Adapted from Feldman (1976) and Wolk (2010)

Fig. 2 Socialization processes and coaching stages

(2) exploration, understanding and interpretation, deals with the moment when the coach questions the coachee so that there is an understanding of the problem situation presented;
(3) expansion takes the problem situation to the scope of possibilities, solutions;
(4) closing ends the session or the process with the coachee's commitment to action.

In Fig. 2, we see the four processes present in the socialization encounter phase that are convergent with the four stages of a coaching process. For socialization to be successful, the four processes must be completed at the encounter stage. In order for coaching to reach its objectives, the four stages must be covered in the sessions and in the process.

The initiation to the task occurs from the assumption of the role by the new employee, advancing to the feeling that he is competent to perform what is expected of him until the moment he starts to be accepted as a full member of the team and the organization (Feldman 1976). Similarly, the first stage of coaching called the introduction/opening serves for the coachee to be introduced to the process and reveal his/her quests in the path of building trust with his/her coach (Wolk 2010). It is believed that both in the initiation to the task of the socialization process and in the introduction/opening of coaching, there is a first feeling of insecurity until the comfort resulting from the feeling of competence and trust is built.

The second process of the socialization encounter phase speaks about group initiation, which is the period of development of acceptance by other members as a team member, requiring skills to establish positive interpersonal relationships (Feldman 1976). On the other hand, the second stage of coaching enters the problem situation of the coachee when seeking to explore, understand and interpret what led him to the process. It is the moment when the coach leads the coachee to a process of reflection on what his goals are, on his understanding of the current situation and a new meaning of what happened. With this, the coachee is made to understand what is happening, to differentiate facts from opinions, and to manage their emotions in the context of the situation. It is a moment of reciprocal study and goal setting (Wolk 2010). There is a conviction that the movements that occur in the process of group initiation with the exploration of reality by the new employee are similar to those performed in the exploration, understanding, and interpretation of coaching.

The third process of the socialization encounter phase addresses the definition of the role by the new employee in relation to the work group, in relation to his tasks and his expected performance. This awareness of his responsibilities favors the new employee to be clear about the relevance and limits of his roles (Feldman 1976). The third stage of coaching portrays the alternatives to change reality through the expansion in which effective actions are projected (Wolk 2010). In this stage, coaching deals with the importance of role performance, showing a clear complementarity with the third process of the socialization encounter phase. Therefore, they are elements that can be used by the use of coaching in the socialization of new employees.

Finally, the fourth process of the socialization encounter phase deals with the importance of a congruent assessment to be carried out between the new employee and the leadership of his group, giving the indicators of his level of progress in organizational life. The congruent assessment allows the new employee to perceive and develop his weaknesses and to value and reinforce his strengths (Feldman 1976). The fourth stage of coaching is aimed at closing the session or process, seeking a process of final reflections on the issues addressed and assuming action commitments for future results (Wolk 2010). In the same way as in previous processes and stages, the complementarity between what occurs in one and the other is present. A congruent assessment of the progress of organizational life that occurs in the fourth socialization process in the encounter phase allows actions to be taken to improve and build the desired future, similar to that explored in the fourth stage of coaching.

Comparatively analyzing the four socialization processes in the encounter phase and the four stages of transformational coaching shown in Fig. 2, the complementarity is analogous to the points of convergence cited between the needs described in the areas covered by the socialization and the benefits offered by coaching shown in Fig. 1. The results presented through the comparisons reinforce the relevance for managers to look at socialization coaching as an additional tool in the activity of welcoming, integrating and socializing new employees, allowing them to express themselves through of their individual characteristics, benefiting the individual and organization in an active movement of seeking to socialize.

Figure 3 describes the path taken by associating theories with what is expected from a socialization process, with the organization's commitment to its own performance and that of the new employee.

HRM is responsible for the process of socializing new employees, notably in the encounter phase, which requires welcoming and integration tactics to be implemented. The areas covered by a socialization program reveal the needs that the process must meet, including choosing the most appropriate tactics for each process. On the other hand, coaching is a personal development tool, often used by HRM with professionals who have an organizational history and often enter conflict zones or present needs for the development of some skills, mostly behavioral. However, research in the existing literature did not reveal the use of coaching in the socialization encounter phase, generating strangeness. It is commented that there are references to the use of coaching in the case of role migration internally, with this movement being

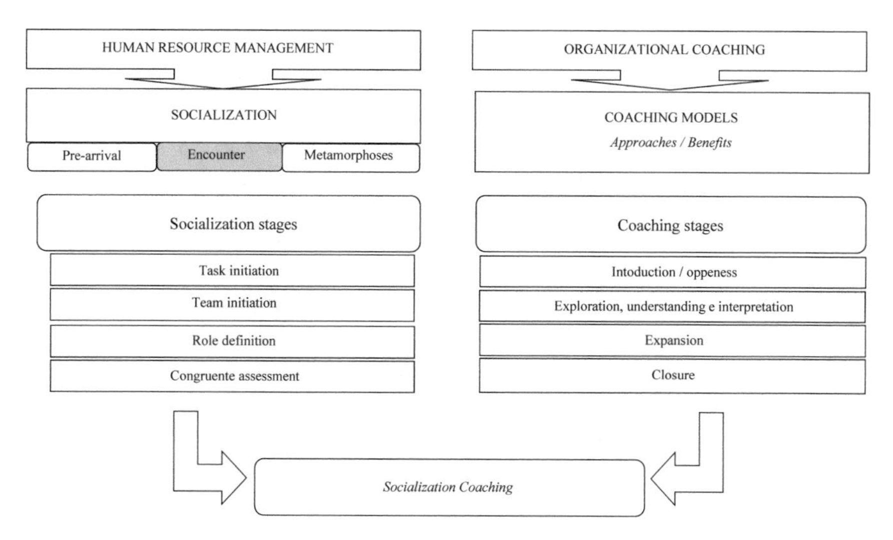

Fig. 3 Socialization Coaching

considered as a new socialization process. References to career transition coaching were also found. Consider that the benefits disclosed as achieved through the use of coaching as a personal development tool are convergent with the needs listed by the HRM for a socialization process to be successful.

According to Barner and Higgins (2007), in the literature, four great coaching models are pointed out, the clinical model, the behavioral model, the systemic model, and the social constructivist model. Although the models are not described here, it is important to highlight that it is understood that the systemic model is the one that can contribute the most so that coaching is effective as a socialization tool. The main search underlying the systems or systemic model is to conduct the process in order to align the coachee's views with that existing in the organizational reality, helping the new employee to adjust to the new system. It is this adjustment that the socialization process aims when receive a new employee, placing coaching as another auxiliary tool. It should be noted that the coachee is also understood as a complete and complex system, analogously to what the organization is considered (Morin 2003). However, the application of coaching in the socialization phase is not only based on the systems model. It also finds support in the behavioral and social constructivist models, albeit with less intensity. Finally, the coach's knowledge of the clinical model can help to identify some pathology, in which case coaching will no longer be the appropriate tool.

The literature also presents us with the main approaches that tend to offer contributions for the application of socialization coaching, with the ontological coaching, positive psychology coaching, NLP coaching, and the developmental cognitive coaching approaches being those with the most connection. Among the countless existing approaches, it is believed that these are the ones that can contribute the most to a coaching process used in the socialization phase, although the other approaches also offer instruments and a theoretical framework that allows to contribute to the process. Coaching approaches are understood as the theoretical assumptions that underpin the practice, revealing concepts and premises about the development of people in an organizational environment or outside it, as well as the difficulties and obstacles encountered in the process and exploring its application dynamics (Cox et al. 2010). It is commented that the approaches tend to offer theoretical and practical contributions for the different models, in a movement of feedback, because they too are more or less linked with the models. Different approaches also tend to permeate the same methodology and the different uses of coaching. Therefore, in the intention of using coaching as a socialization tool, the focus should be on the needs displayed by management for what is expected to happen in the encounter phase of the new employee with the organization, with one or the other approach being the most adequate, depending on the organizational reality and the coach's affinities.

Finally, Fig. 2 shows the similarity between the processes present in the encounter phase of socialization (Feldman 1976) and the steps considered in the transformational coaching methodology (Wolk 2010), used for the proposal. It should be noted that other methodologies, schools, or types of coaching may have a different division of the stages. However the complementarity between the processes of the socialization encounter phase and the structure of a coaching process remains. It is noteworthy that both the socialization process in the encounter phase and the coaching process distributed between its sessions are not airtight and happen linearly. The socialization process in the encounter phase does not occur by progressing neatly from the initiation to the task, moving on to the group initiation, progressing to the definition of the role, and ending with a congruent assessment. The beginning occurs with the arrival at the organization, but the processes can move forward and backward in a non-linear movement. The initiation to the task may have been successful, as well as the definition of the roles, however the initiation to the group may not have been as effective as the other processes. In other circumstances, the group initiation may have been good, while the task initiation was not, depending on a congruent assessment to follow the process. The same phenomenon occurs in the coaching process with its steps within the programmed process and with the same steps happening within a session. It is the non-linear movement that allows that the socialization process and coaching as a socialization tool can be effective, because they are not radical, restrictive, or definitive. It is understood that both work with the alternatives.

It appears from the analysis of the literature on socialization and coaching, that this, being used as a socialization tactic, will be classified as individual in relation to the context, since it is an individual application to encourage the new employee to express its heterogeneity (Jones 1986). It is about avoiding the level common in institutionalized and collective tactics that tend to homogenize individual behaviors (Herrmann 2013). For this reason, according to the author, the practice is also aligned with the use of individualized tactics in which the new employee assumes the role of his insertion in the organization, playing an active role as envisaged in coaching. Coaching is a personal development tool that considers and stimulates the self-learning process (Anderson et al. 2008; Whitmore 2002) and can encourage the new employee to develop the self-learning process directed to the needs found in the socialization process (Ellinger 2004). With this, the new employee will be able to give his contribution in a creative, proactive, and innovative way as a unique and singular member. Still in relation to the context, socialization coaching can be considered an informal tactic due to the characteristic of coaching as a non-formal process in which the employee is led to understand and assume his role, although it also has characteristics derived from the formality that is the separation from the other collaborators to develop the competencies of the function (Jones 1986).

Jones (1986) places the sequential and fixed tactics linked to institutionalized tactics in the content group and the non-sequential and variable tactics linked to individualized tactics. With regard to content tactics, socialization coaching is strongly in tune with individualized tactics because it does not display a schedule for carrying out tasks and neither does it have information about the path to be followed by the new employee. Again, socialization coaching aligns with Herrmann's (2013) thinking as an element to stimulate the initiative and with Ellinger (2004) to lead the new employee to the directed self-learning movement, just as it is in the nature of coaching as a development tool folks.

The group of social aspects proposed by Jones (1986) includes Serial and investiture tactics on the side of institutionalized tactics and disjunctive and stripping tactics on the side of individualized tactics. There is an understanding here that socialization coaching is aligned with individualized disjunctive tactics and institutionalized Investiture tactics, in opposition to the classification of tactics in the social aspect. However, this association is convergent with the essence of coaching in order to lead the new employee to understand and interpret in order to build his own definitions of each of the challenges that the new role requires (Gallwey 2000), characterizing itself as an individualized disjunctive tactic. Socialization coaching is also convergent with institutionalized investing tactics that work with the concept of reinforcing personal beliefs in their abilities and skills. It is understood that this alignment of socialization coaching finds support in the propositions of Herrmann (2013) to take the individual to play an active role in his socialization process, revealing his proactive, creative, and innovative side. As well as in the theories of self-learning directed towards the interests of the individual and that are in line with organizational purposes (Ellinger 2004).

Figure 4 outlines the four socialization processes in the encounter phase (Feldman 1976) and the four stages of transformational coaching (Wolk 2010) used to support the application of socialization coaching and inserts the four moments of the new employee: the I Am (as understanding), the I Am (as identity), the I Create, and the I Manifest (Wilber 1997, 2011). It is intended to show the manager that not only the individual must have a systemic view of the organization, but it is up to the organization to understand, receive, and welcome the individual as a complete, complex, and interdependent system so that he can and wants to integrate without homogenizing. It is the point outside the curve that will generate organizational competitive advantage.

When looking at Fig. 4, it can be seen that socialization coaching works from a perspective of movement and exchange that occurs between the parties involved in the process:

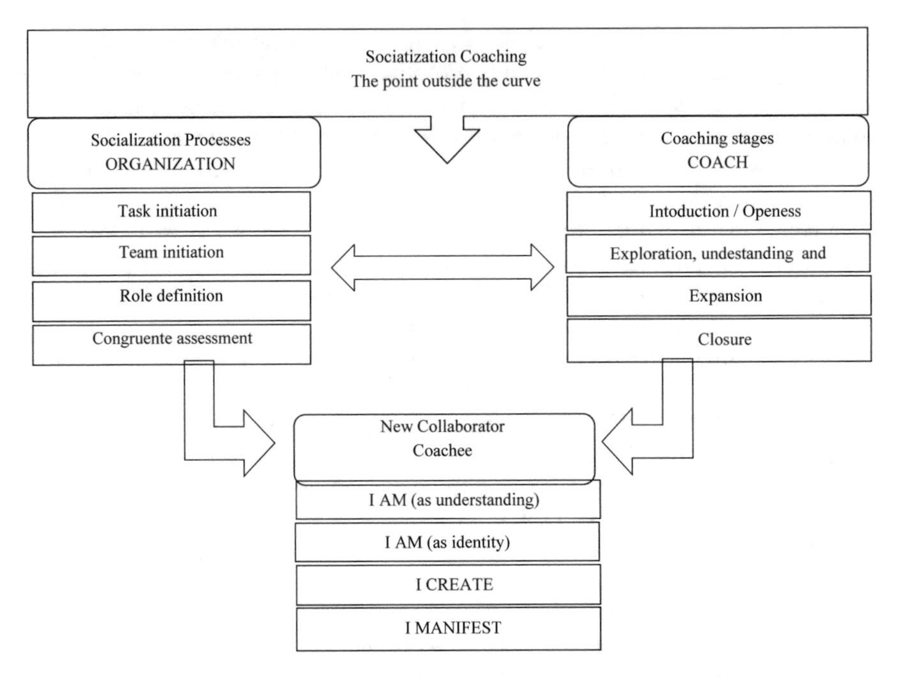

Fig. 4 Interdependencies of socialization coaching. *Source* Adapted from Feldman (1976), Wilber (1997, 2011) e Wolk (2010)

(1) the coachee, who is the new employee for whom socialization is directed;
(2) the organization, which is responsible for the socialization processes; and
(3) the coach, who is responsible for applying coaching as a socialization tactic that contributes to the active role of the new employee in the process.

Socialization coaching starts from understanding the needs of the new employee to be situated in the new environment. The *I am (as understanding)* allows him to understand himself as a new employee in the face of the task, the group, and the role, considering a congruent assessment that the coach will be able to explore when opening the coaching process. The *I am (as identity)* gives the dimension of an identity to the task, the group, and the role through a congruent assessment conducted by the coach in a movement of exploration, understanding and interpretation of the new reality. The *I create* advances towards the perception of the contribution to the task, the group, and the role in a movement to expand alternatives using an evaluation process brought by the coach. The *I manifest* allows the new employee to evaluate himself in a congruent way as he expresses himself through the task, in the group and in the role or roles he plays. However, this is a circular, unidirectional and non-linear movement.

There is a conviction that the new employee tends to evolve the level of perception about the task, the group, the role, and performance asymmetrically. Sometimes, a reassessment allows the new employee to have a new perspective that will allow him to be more effective in what he wants and what the organization needs. In this way, just as the *I Am (as understanding)* affects the task, the group, the role, and the assessment; the task, the group, the role can affect the *I Am (as understanding)*. The same movement occurs with the *I Am (as identity)*, the *I Create* and the *I Manifest* affect and be affected by the task, the group, the role, and the evaluation. And the perceptions can vary within a socialization coaching stage in which the conditions of confidence generation, exploration, expansion, and closure are created for each of the situations faced by the new employee, as well as within the sessions, the steps, the stages, and the process in its completeness.

It is in this variability, unpredictability, multiplicity, and singularity that socialization coaching is understood to be aligned with the concern of not transforming all employees into a common mass, eliminating the individual characteristics that led the organization to hire him (Herrmann 2013). Social coaching aims to encourage the new employee to develop in the task, to join the group, to assume the role, and to evaluate himself congruently so that he maintains his identity offering the organization his creativity, his proactivity and its capacity for innovation (Allen 2006; Cooper-thomas and Anderson 2006).

Thus, the intention is that socialization coaching serves as an effective tactic that can deliver an effective socialization process for the new employee to his new organization, understanding them as complex, complete, and interdependent systems (Morin 2003).

4 The Socialization Coach That Welcomes and Accepts People Without Homogenizing

It is understood that the socialization coaching should be conducted by a professional coach who has the required training, in addition to knowing the duties of the organization's HR area. It is believed that to work with socialization coaching, the coach must have knowledge about the Recruitment and Selection processes, the Compensation and Benefits packages, the list of Positions and Salaries, the Performance Assessment indicators and the Training and Development programs, among other attributions in the HR area, including welcoming and integration tactics for the socialization of new employees. This concern is justified because it is a coaching process with clear objectives for the parties involved within what is expected for socialization to be positive. However, it is noteworthy that the coaching process as a personal development tool can advance to areas of the coachee that, apparently, go beyond the scope of socialization. Coaching as a process, besides being oriented towards socialization, is oriented towards the individual, who is a complete and complex system with all its interdependencies (Wilber 1997; Dass and Parker 1999; Morin 2003). For pathologies the path tends to be psychotherapy or psychoanalysis.

Particularly, it is suggested that the professional who acts as a socialization coach should know the organization in which he works, preferably having links with it. It is known that one of the prerogatives of coaching as a process is its non-directness and the non-existence of expectations on the part of the coach regarding the coachee, however, the previous recommendation is made because the socialization coaching is located within a delimited area. There is an expectation of the coachee and the organization in order that at the end of a certain period the new employee is considered a full organizational member, developing his tasks, being part of the group, assuming his roles and exhibiting the expected results in accordance with his competences. It should be highlighted here that expectation, as well as directivity, is not the coach's roles, however, when knowing his area of expertise the coach will be able to measure how far his role and responsibilities go (Gallwey 2000).

Thus, a socialization coach must have the traditional training of a coach, being up to each organization to define its level of excellence to take charge of the socialization process for new employees. The suggestion that a socialization coach has links with the organization in which the new employee is seeking to become a full member is not exclusive. From the analysis of the bibliographic research carried out and relying on the experience of those who produce the present proposal, it is understood that this is an option. However, there is nothing to prevent an organization from hiring independent coaches to carry out the socialization coaching processes, once the objectives are well defined and the fields of action of each party are defined.

It is believed that the socialization coach should deepen the knowledge of what is meant by adult learning. Thus, it is up to the socialization coach to have a solid base of the adult learning process, knowing Knowles' andragogy of the 1970s (Knowles et al. 2005); the experiential learning proposed by David Kolb in 1984 (Kolb 2015); and Mezirow's transformative learning from 1990 (Cox et al. 2010). Finally, the coach must know and be aware that socialization coaching is a self-directed learning process (Ellinger 2004) in which the coachee, the new employee, is the focus.

From the next section, an application structure for socialization coaching for new employees will be presented.

5 The Application of Coaching to Welcome and Integrate Without Homogenizing

The use and application of coaching for a given area can use resources from different models, philosophies from various approaches and structures from different methodologies, depending on the education, knowledge, capacity, skills, and the identification of the coach with this universe. Meanwhile, the application and use of socialization coaching presented here are built from the perspective of the researcher who uses the transformational coaching methodology to achieve the objective of what socialization through welcoming and integration programs intends. The choice is made according to the alignment among the beliefs, values, and training of those who propose the application, and these sources may vary according to who the socialization coach is. However, it is believed that the common point to all socialization coaching applications are the expected results of the process: making the new employee a full organizational member.

According to Taormina and Law (2000), actions designed for the socialization process can focus on the individual to develop their interpersonal skills, self-management skills, and psychological preparation of the new employee. Likewise, the authors emphasize that actions with a focus on the organizational environment can be thought of, such as understanding the organizational culture, training to improve technical skills, encouraging the support of co-workers, and projections of the future in the organization. Socialization coaching is an individualized tactic with an individual and informal character in terms of context and a non-sequential and variable character in terms of content; and with regard to social aspects, it is an institutionalized investiture tactic and a disjunctive individualized tactic. Like other tactics, socialization coaching aims to make the new employee feel part of the new organization (Cooper-Thomas and Anderson 2002; Madlock and Chory 2014). In this way, socialization coaching will have fulfilled its role when it contributes to increase the involvement, satisfaction, and commitment to work and the organization and to minimize the intention of premature departure of the new worker. In the same way, it can be assessed that the socialization coaching has fulfilled its role, when it contributes so that the new employee is able to reduce the feelings of ambiguity regarding his task, with the integration in the work group, regarding his role in the team and in the organization, combining these objectives with a congruent and positive assessment of his performance.

Therefore, the socialization coach must structure a coherent process with appropriate questions, tools, and procedures that guide the new employee on the chosen path, in line with organizational objectives (Lane and Corrie 2007).

It can be said that the methodology that the socialization coach uses tends to be the map, but it is not the territory, making an analogy with the sentence said by Alfred Korzybsky in 1933. It defends the use of the socialization coaching methodology as a flexible proposal that allows the coach to have a map in hands, but it is up to the coachee to know its starting point and determine the arrival point. The path will be built in the interaction between coach and coachee.

The socialization coach must remember that it is each new employee who will give meaning to each situation experienced (Boose 1984), be it in his initiation to the task, in his initiation to the group, in his definition of the role or his evaluation in the organization where he now seeks to become a full member. It is important to highlight that each new employee will assimilate the contents or direct his learning in accordance with his principles, values, and worldview, since it is understood that the external reality is the same for everyone and what changes are the perspective of whoever observes it (Wolk 2010). It turns to the importance that the manager, HR area and coach understand that people are a point outside the curve, so they need to be recognized so that they feel welcomed and integrate without homogenizing.

The transformational coaching methodology that will be used for the application of socialization coaching, assumes that the world is the expression of the observer, and may vary according to the individual interpretation (Wolk 2010). With this in mind, the socialization coach will have the mission of helping the new employee to expand his vision to be a different observer based on transformative learning. It is important that the socialization coach has implicit in its approach the reminder that each individual is part of different life systems that encompass the self, such as personal relationship, family, friends, work, and community (McLean 2012). However, here the focus is on work in the dimensions of the task, the group, the role, and performance, with the awareness that the other systems are always present. It is not possible to simply fragment the individual, because he is a system within another system (Morin 2003). It is understood that at work it is important that personal relationships are healthy, that new friendships are built, and that people get involved with the community that is affected by the organization, leading the new employee to become a full member of the organization.

Wolk (2010) describes the transformational coaching process as being divided into four stages and with seven different steps. For each stage there are one or more corresponding steps, as follows:

Stage I: Introduction/opening—Step 1.
Stage II: Exploration, understanding and interpretation—Steps 2, 3, and 4.
Stage III: Expansion—Steps 5 and 6.
Stage IV: Closing—Step 7.

Next, the stages, steps, and suggested questions based on the transformational coaching methodology (Wolk 2010) for the application of socialization coaching will be presented. In the tables elaborated, the conventional questions of transformational coaching will be described in the first column, while suggested questions for socialization coaching will be listed in the second column. After each table, comments will be made that can help the socialization coach to build his own approach during the sessions.

6 Stage I: Introduction, Opening, and Generation of Trust

The first meeting is essential for a coaching process to be successful (O'Connor and Lages 2002). In socialization coaching there will be no different, because it is the moment when the new employee seeks help of his own free will or at the suggestion of the HRM. Thus, the coach must be attentive, because it is in the first contact that the norms of the process that are initiated are established, that a contract is formulated, that the trust is generated or not, and that the situation is contextualized.

6.1 *Step 1: The Introduction, The Opening, and The Generation of Trust*

Step 1 is composed of three moments: the introduction, the opening, and the generation of trust (Wolk 2010).

The introduction refers to the moment of arrival when the new employee and the coach introduce each other. It is important that the environment in which the socialization coach receives the new employee is prepared so that he can feel comfortable and safe. The socialization coach must take care to not be interrupted and to not allow interference by bosses, managers, or directors. Here, the operating rules for the session that begins and the sessions that will be held in sequence are also clarified, establishing the commitment contract between the parties. It can be a real contract with terms and conditions, as well as a verbal contract in which the parties make their commitments. In addition, the coach must pay attention to the commitment to confidentiality on the topics covered in the sessions, since the issues addressed between the new employee and the socialization coach must remain between them. The socialization coach's job, even though he is a contractor of the organization, is not to collect information to justify a termination, since he is not a superior of the new employee. The role of the socialization coach is to contribute so that the new employee develops for his own benefit and that of the organization, based on building a relationship of trust with the new employee. This is the time for the deep detailing of the conditions required for socialization coaching to be effective and this occurs in the first session of the coaching process. In the other sessions, the introduction is the protocol moment to start the work, with the presentation of the demands of the new employee.

The opening follows the introduction in which the session's working context is generated, at which point the new employee begins to reveal his search. The socialization coach should not be in a hurry to ask the questions he considers important, because the new employee will demonstrate when he will be ready to hear them. For this reason, the generation of the context is an important moment in the relationship that possibly will last for three to 4 months.

The socialization coach must exhibit tranquility, empathy, and positivity so that the new employee feels safe in the environment in which he finds himself and, mainly, in the process in which he will be the focus.

It is worth remembering that the socialization coach does not have the role of judging or evaluating the new employee, but rather of building a relationship of trust so that he can fully develop his potential in relation to the task, the work group, the organizational role, and results. In line with this, the contribution of socialization coaching so that the new employee reduces his internal and external conflicts, increases his organizational commitment, and has a high degree of job satisfaction. For this reason, it is important that the socialization coach has deep knowledge about what the organization develops in welcoming and integration programs through the learning trails, or another name that represents the programs, to benefit the new employee on his/her journey to become a full organizational element.

Therefore, socialization coaching needs to be well thought by the coach so that this first stage, with its first step and its three different moments (Introduction, Openness and Trust) is successful. It will be from this initial success that the socialization coach will be allowed to enter the world of the new employee to ask the questions that are important and use the other resources of personal development that he deems relevant to achieve the established goals.

In Fig. 5 below, the questions in the first column are those suggested by Wolk (2010) for Stage I and Step 1 of a coaching process. In the second column, comments

<table>
<tr><th>TransformationalCoaching</th><th>Socialization Coaching: comments and aplications</th></tr>
<tr><td>What place or position do you occupy in the company?</td><td rowspan="6">The questions used in a Transformational Coaching process will not always be indicated for Socialization Coaching. From the suggestions in column one, the appropriate questions are considered:
▪ What position do you occupy in the company?
▪ What do you do in your task?
▪ How do you feel about her?
▪ What are your concerns?
The Socialization Coach probably knows what the new employee's occupation is, however it is important that he verbalizes the answer so that the Coaching process begins.
At this point other questions are suggested that pose the other dimensions of socialization in the coaching session:
▪ How is your relationship with the working group?
▪ What is your role in the group and organization?
▪ What is your performance evaluation?
To complement, it is suggested that the questions about how the new employee feels and what his / her concerns are, focus on each of the four items of socialization.
It is important to highlight that the Socialization Coach should not have expectations regarding his coachee, knowing, however, that he will have expectations regarding the coach and the process.</td></tr>
<tr><td>What do you do in your task?</td></tr>
<tr><td>How do you feel about her?</td></tr>
<tr><td>Family, studies, free time?</td></tr>
<tr><td>What are your concerns, your interests?</td></tr>
<tr><td>What experiences have been important in your life and why?</td></tr>
</table>

Fig. 5 First Stage and Step 1 of Socialization Coaching

are made about the possibilities of use to the purpose of the new employee actively seeks socialization.

The questions maintained for socialization coaching originating from the transformational coaching methodology are justified, because they directly meet the expected objectives. These are questions that lead to situations that may be embarrassing or limiting the new employee on his way to becoming a full organizational member.

The socialization coach must remember that it is the first meeting of a process that can span up to twelve sessions. Probably, the new employee will be loaded with uncertainties and insecurities regarding the task, the group, the role, and how he evaluates himself and is being evaluated. Therefore, it is believed that the session should also focus on how the new employee feels about the I am (as understanding), the I am (as identity), the I create, and the I Manifest. Thus, in addition to the suggested and commented questions, the socialization coach should explore a little of the new collaborator's history and trajectory to understand how he feels, how he is, how he creates, and how he manifests himself in the organization from his own eyesight. It is important for the new employee to share his experience, because in general people like its own story and asking about it is a sign of respect for the new employee and an opportunity for confidence and learning for the coach. The coach may also ask, *How did you get here?* The question is about what was the new employee's way to find the vacancy, how was the selection process, and what he thought and thinks about the organization. These questions enable the coach to identify any sign of breach or violation of a psychological contract established by the new employee with the organization in the process of anticipatory socialization and in confronting the organizational reality in his first contacts (Allen 2006; Robinson and Rousseau 1994).

As it is the initial phase of the meeting phase of the socialization of the new employee, it is understood that he is not yet fully integrated into the organization with regard to the task, the group, the role and that perhaps his level of satisfaction with the work is below expected, as well as his performance. In addition, it should always be remembered that the new employee is a complete and complex system that is part of other systems, such as the family system, the system of friendships, and the system of a community in addition to the organizational system (McLean 2012; Morin 2003). That is why it is important for the socialization coach to keep in mind that the way the new employee IS (as understanding) at the moment has repercussions on what he IS (as identity) in the organization and that has consequences on the way he CREATES and the way he MANIFESTS in it (Wilber 1997). Therefore, the coach must welcome the coachee in order to offer an environment of integration, freedom, and trust so that the new employee has the motivation to share the best of himself in the organization (Wolk 2010). It is important for the organization that he is accepted and welcomed as a point outside the curve to integrate without homogenizing.

To conclude the session, the socialization coach must be aware that he has completed the first stage and the first step of the coaching process in which it was treated the introduction, the opening and the generation of trust with the new employee. Some questions can be suggested to close the session, *What did you learn today? What is your commitment to action? How will you notify me about the evolution?* These are questions that commit the new employee with his effective participation in the search for socialization. The coach must have led the new employee to explore, understand and interpret, as well as expand and end the session in relation to the first stage of the coaching process, considering the interdependence with the different dimensions of himself and socialization. It is understood that this is how coaching develops, always opening, and closing a process within another process.

7 Stage II: Exploration, Understanding and Interpretation—Steps 2, 3, and 4

From the second session, normally, the second stage begins, involving Steps 2, 3, and 4. However, the sessions do not necessarily coincide with the steps. It can happen, and it will happen, that Step 2, for example, occupies two sessions, or that Step 3 equally occupies two other sessions, while Step 4 is completed within one of the sessions that is also occupied by Step 3. Transformational coaching is not a linear process, nor will socialization coaching be.

It is believed that in socialization coaching, the second stage is marked by the discovery of the objectives that the new employee hopes to achieve when exploring the problem situation in order to seek to understand it, interpret it and reinterpret it in order to be able to reframe it, generating new perspectives within the organizational environment. Through socialization coaching, the new employee will be led to take a stand against the task, the group, the role, and the performance in a process of confronting perception with facts.

7.1 Step 2: The Demands Of The New Employee

Thus, as occurred in the first session, the process begins with the introduction and the opening, understanding that trust has already advanced one step. The new employee begins by presenting his demands so that the socialization coach can explore them, leading him to understand them and to interpret and reinterpret them. Here the highlighted questions should carry with them the ability to make the new employee reflect reflexively on the situation under analysis, because delving into the situation that proves difficult will allow him to understand the reflexes that it has in his goals aligned with organizational and group objectives (O'Connor and Lages 2002; Wolk 2010). The short, medium, and long-term individual goals analyzed in their relationship with the tasks, the group, the role, and performance are reflected in job satisfaction, in the commitment to the organization, and in the existence of conflicts (Madlock and Chory 2014).

Therefore, once again, the importance of the socialization coach getting to know the welcoming and institutional integration program is highlighted, in addition to other practices to be able to associate the expectations of the new employee with what is measured in the organization. It is important that the socialization coach knows which are the main indicators used by the HR area, such as the employee satisfaction index, the salary range by function, the number of employees by area, absenteeism rates, the average promotion times, the voluntary or involuntary turnover, the cost of hiring, the quality of hiring, the minimum financial return required, the return from training, among other indexes used in the organization.

In the socialization coaching process, Step 2 focuses on identifying the problem situation of the new employee, seeking to understand how he is where he is; who he is where he is; what he creates where he is; and how it manifests itself where he is (Feldman 1976; Wilber 1997; Wolk 2010). Bearing this in mind, the socialization coach, using the appropriate questions, allows the new employee to locate himself in the organization and in himself to interpret and reinterpret the problem situation, creating alternatives to set the goals that lead him to achieve the goals.

In Fig. 6 below, the questions in the first column are suggested by Wolk (2010) for Stage II and Step 2 of a transformational coaching process. The second column provides comments on how the questions can be used in socialization coaching.

The questions of socialization coaching, based on transformational coaching, should accompany the moment experienced by the new employee and witnessed by the socialization coach. Often, the new employee is in a situation of tension caused, much more, by internal factors than external to him (Sieler 2010). The subtlety of the questions asked by the socialization coach should stimulate, provoke, and lead the new employee to become aware of the general picture in which he finds himself in relation to what is expected of him, as well as in relation to what he has of expectations for the organization. Therefore, the questions must consider the reality perceived by the new employee in relation to how he feels he is, that he is, that he creates, and that he manifests in relation to the task, the group, the role, and the results (Feldman 1976; Wilber 1997; Wolk 2010).

It is understood that Step 2 is a crucial moment in socialization coaching, because it deals with the perception of a reality that is loaded with emotions that can distort what is in confrontation with what the new employee perceives it to be.

Thus, with the intention of completing Step 2 and preparing the work for Step 3, it is suggested to adapt the tool called "The Wheel of Life" using the dimensions involved in the work. It is up to the coach to adjust the tool, allowing the new employee to evaluate himself within each part of his "Wheel of Work".

Therefore, the socialization coach must think that the new employee is at the stage of defining how he feels in relation to how he is, who he is, what he creates and how he manifests himself in the task, the group and the organization (Wilber 1997; Wolk 2010).

<table>
<tr><th>Transformational Coaching</th><th>Socialization Coaching: Comments and adaptations</th></tr>
<tr><td>What is happening?</td><td>What is happening? it is an objective question that the Socialization Coach can address to the new employee. It is understood that someone who has a certain problem, sometimes, does not question himself truly and reflexively about what appears to be a problem. The encounter phase of the new employee is a moment when he is in the zone of discomfort and, often, the ability to perceive objective reality is clouded by feelings of insecurity and anxiety. The discomfort sometimes prevents the new employee from having a real and systemic view of the situation. When confronted with the question, the new collaborator will have to verbalize an answer, starting to redraw the mental map and creating possibilities by expanding the vision and envisioning alternative solutions.</td></tr>
<tr><td>What would you be interested in working on?</td><td>What would you be interested in working on? is a typical question in the Coaching process and it also applies to Socialization Coaching, because just as the previous question leads the new employee to verbalize his current situation within the organization, the work group and the task, enabling the interpretation of reality and the emergence of alternatives.</td></tr>
<tr><td>What do you want to happen with this Coaching?</td><td rowspan="3">▪ What do you want to happen with the situation?
▪ What are your expectations regarding the situation?
▪ What results do you expect for the situation?
These are fundamental questions in a socialization process, since the objective is to transform a previously external person into a full member of the organization. With the socialization process in progress, the questions proposed here lead the new employee to take an active role in his quest to become a full member of the organization and the closest working group. The Coach must understand that a fundamental step towards reaching the general objective of the new collaborator being a full organizational member is that he be accepted into his closest working group. Therefore, the facilitating questions presented are intended for the new employee to be able to see the big picture to rescue his best abilities that can lead him to the desired destination with high performance, role definition, group acceptance and tasks performed competently. The questions stimulate logical, critical and analytical reasoning, which is sometimes clouded by the apparent complexity of the discomfort experienced by the new employee.</td></tr>
<tr><td>What are your expectations?</td></tr>
<tr><td>What are the expected results?</td></tr>
</table>

Fig. 6 Second Stage and Step 2 of Socialization Coaching

7.2 *Step 3: The Present Situation*

Step 3 uses the information collected in the previous coaching sessions to deepen the understanding of the current situation, in order to differentiate what the coachee perceives as reality from what is reality. Likewise, in socialization coaching, the coach will use the data and information obtained in previous sessions to make the new employee identify the differences between his perceptions and the external environment. With this differentiation it is possible to measure more accurately whether the situation is a problem or just seems to be a problem, making it possible for the new employee to channel his energies, motivation, and skills to act in a congruent way in the pursuit of his goals. The socialization coach leads the new employee to differentiate what is fact or merely opinion in his organizational reality, in relation to the task, the group, the definition of the role and the self-assessment of performance. With this, the new employee can redefine what he pointed out as a source of conflicts that may have influenced his own assessment of satisfaction or commitment at work.

It is up to the socialization coach to have the knowledge and skills necessary to conduct the process to align the individual's purpose, vision, values, and objectives with the mission, vision, values, and organizational objectives.

In the socialization process, Step 3 is essential and can be more extensive, because the new employee tends to be in an area of discomfort due to the environment, work, and people in the new organization, in addition to feeling questioned about his ability to see the world as it really is (Allen 2006). It is noteworthy that the ability of the socialization coach makes a difference in the new employee's motivation to adopt new postures in the face of a reality that is under construction (Whitmore 2009).

Step 3 of the second stage of socialization coaching tends to be a defining moment for the individual's goals and objectives to be achieved in line with those of the organization (Saks et al. 2007). These are all questions that lead the new employee to explore situations in depth and can bring up emotions that involve people from the internal and external environment (McLean 2012) (Fig. 7).

Here it goes back to the point that it is relevant for the socialization coach to see the person behind the new employee (McLean 2012). Deepen in the exploration of the characteristics and pretensions of the individual who is a complete, complex and interdependent system (Morin 2003), which integrates other complete and complex systems such as the family, friends, community, and intimate relational system, in addition to the organizational system (Wilber 1997, 2011; McLean 2012).

Therefore, it is possible to propose a specific work with the purpose of making the new employee aware of the mission, vision, values, and organizational objectives and discover the purpose, vision, values, and individual objectives.

One begins to move towards the formulation of goals for each part of the new employee's organizational life, following the socialization processes planned in line with the other parts of the work environment involved in the socialization process. The exploration of the current situation in Step 3 goes beyond the perception of the new employee in relation to the external environment, because the present step leads

Transformational Coaching	Socialization Coaching: comments and adaptations
What is happening? What is the situation? Where does this happen?	Step 3 starts again with the question, *What is going on?* that may have variations, among them, *What is the situation?* They are recurring questions in a conventional Coaching process and applicable in Socialization Coaching, since they are exploratory and reflective. The Socialization Coach must have the ability and the sensitivity to apply them at different moments, allowing the new employee to answer them from a different perspective related to the task, the group, the role and the performance. In addition to using back tracking techniques (coaches understand), the Socialization Coach can amend to the received answer the question, *Where and when does this happen?* so that the new employee brings the situation to a time and place. This allows him to differentiate what an opinion or judgment is, minimizing some level of dissatisfaction and maximizing involvement with the organization. The Socialization Coach asks the questions to give the new employee the opportunity to deepen the analysis of his perceptions, seeking to take him to a logical, self-critical and analytical process that tends to create new solutions to real problems. We seek to awaken in the new collaborator the expansion of his interpretations so that he is able to understand the organization as a complete and complex system, while it is up to the Socialization Coach to do the same with him.
Who are the protagonists?	The question, *Who are the protagonists?* of the problem situation brought about by the new employee can reveal where it personifies itself. The Socialization Coach must be careful not to look just like an interloper, because it causes the new employee to initially turn his attention to other actors, internal or external to the organization. However, the goal of the Socialization Coach is to hold the new employee accountable for articulating his work in the task and in the working groups that tend to favor the definition of the role and performance by building synergies.
What would you like to happen or have happened?	
What do you need to happen to feel at peace?	Other questions are suggested in Transformational Coaching

Fig. 7 Second Stage and Step 3 of Socialization Coaching

What would change if 'X' happened? Do you have any idea why it happened what happened? Or, what is your opinion on why 'X' happens? What makes you think that ...?	and that can be used in Socialization Coaching, safeguarding the goals of the new employee and the organization, among them, *What would you like to have happened?* and *What do you need to happen?* tend to reveal solutions. The questions, *What would change if 'X' happened ?, Do you have any idea why it happened what happened?* or, *What is your opinion about why 'X' happens ?*, are questions of deepening the problem situation, being the Socialization Coach responsible for using them in accordance with the objectives of socialization for the creation of new scenarios regarding the task , the group, the organizational identity and performance. With these questions, the new employee is encouraged to be proactive, creative and also, innovative to articulate interpersonal relationships in order to reach the goals of integrating the organization as a full member.
What prevents you from taking action? How do you feel about that?	One of the coaching's most provocative and powerful questions that also applies to Social Coaching, *What prevents you from taking action?* can and should be used by the Socialization Coach. However, he must be aware that it can trigger contact with the emotions of the new employee or also transfer responsibility to people or situations beyond his control. Sometimes the paralysis of a new employee is related to fear and insecurity in the presence of his new co-workers who are full organizational members. For this reason, the subtlety and sensitivity of those who ask questions are required to expand the alternatives of the new employee and not to restrict them. The Socialization Coach needs to know that, sometimes, the question can cause the rejection of the new employee, because it questions his ability to act, leading him to close in a movement contrary to the intended.
What's in your left column (hidden thoughts and feelings)?	The question *What's in your left column?* can be translated as *What do you think you haven't said yet?* about the problem situation at that moment of its socialization. It is a technique that seeks to unravel the internal dialogues, which are often negative, and which may not always, in a good tone, verbalize them. However, they exist and cannot be disregarded. Sometimes, in the midst of the whirlwind of unspoken thoughts, there are some that can be the solution to the problems that the new employee has with regard to the task, the group, the role or performance. That is why it is up to the
Why don't you express them?	
What are the consequences (desired and unwanted) of	

Fig. 7 (continued)

<table>
<tr><td>saying and not saying this?</td><td>Socialization Coach to explore these thoughts by asking questions, Why not express them? or What would be the consequences of saying or not saying? It can still be said that they are exploratory questions with a high potential for creating new scenarios. Some thoughts that cannot be said as they appear in the mind of the new employee can make a major contribution to the organization and to himself, after being re-articulated and adjusted to the organizational reality based on respect for himself, the group and the organization. This rearticulation carried out by the new employee will require him to develop communicational and relational skills that can help him to occupy his space as a full member based on his action plans.</td></tr>
<tr><td>On what data / observations is your opinion based?</td><td rowspan="5"><ul><li>What is your opinion of the situation based on?</li><li>What makes you think your opinion is valid?</li><li>How do you reach this conclusion from this data for the situation?</li><li>What is the responsibility / concern / matter of what is happening?</li><li>Could there be opinions that are being taken as facts?</li></ul>These are all exploration questions that lead the new employee to take a concrete position in relation to his demands. The questions asked by the Socialization Coach are not intended to contradict or doubt the new employee, so they should be asked respectfully in order to once again differentiate facts from opinions, observations from judgments. Facts are important to everyone, as are observations, however opinions are not always relevant, just as judgments rarely contribute to the benefit of themselves or the organization. Therefore, the ability of the Socialization Coach must have a solid foundation of knowledge of the Coaching processes, as well as the HRM structure of the organization in which he works. Even so, it is believed that a fact is loaded with assumptions from those who interpret it, which can vary between people. Thus, it highlights the importance of exploratory questions that tend to deepen the understanding of the new employee about his organizational reality with regard to the moment he lives in his search to be a full member in relation to the task, the group, the role and the performance.</td></tr>
<tr><td>What makes you think your opinion is valid?</td></tr>
<tr><td>How do you come to that conclusion from these data?</td></tr>
<tr><td>What is the responsibility / concern / imposition of what is happening?</td></tr>
<tr><td>What opinions are you taking as facts?</td></tr>
</table>

Fig. 7 (continued)

	Following are some comments on the questions, *What opinions are you taking as facts?* and *What makes you think your opinion is valid?* These are two important questions to be asked in the socialization process, although again it requires sensitivity, care and self-interest on the part of the Socialization Coach so that they do not sound aggressive and disrespectful. The questions tend to provide the opportunity for the new employee to validate himself / herself against the work group, helping them to take a position on the definition of the role. Finally, the suggested question, *What does it matter what is happening?* can suffer variations, but it is useful for the Socialization Coach to evaluate how the new employee sees himself in the issue of organizational commitment.
Which commitments were broken or not honored? What excuses were not offered (from both the coachee and a third party)?	▪ *What commitments were broken or were not honored in the situation?* ▪ *What excuses have not been offered?* We reach to another crucial point of socialization in the encounter phase in which Socialization Coaching has a lot to contribute through the two questions above. The psychological contract established by the new employee in relation to the organization does not depend on the organization's will, but it is a reality. On the other hand, it is up to the organization to confirm it or not, even if it has not been signed by it. The new employee enters the organization with his expectations based on his individual beliefs and created during the anticipatory socialization phase. Failure to comply with this contract has generated a high turnover rate in the phase of encounter of the new employees' socialization. And Socialization Coaching, when asking about the commitments that the new employee understands that have not been fulfilled, notably by the organization, in his view, even in the figure of co-workers, manager, director or by the work group, opens the possibility that the psychological contract is not broken, violated or that it can be reinstated. The Wheel of Work created by the new employee in Step 2 can give clues about his position on the issue, competing upon the Socialization Coach to explore the situation in time to correct the likely breaches of the psychological contract . The second question about excuses not given goes in the

Fig. 7 (continued)

	same direction and the Socialization Coach can use it to assess situations in which HRM can intervene or modify its welcoming and integration planning so that the socialization process is effective, ensuring the organizational investments made in recruiting and selecting, training and developing our new employees. These are deep questions that demand from the Socialization Coach knowledge, skill and responsibility supported by the trust generated between the parties since the first step of the Coaching process.

Fig. 7 (continued)

to a deep exploration of the internal environment so that there is an understanding of its interdependence (McLean 2012; Wilber 1997).

We move on to the concept that coaching sessions are processes within other processes in an evolutionary movement until reaching the final goal that the new employee feels socialized because he decided to socialize (Wilber 1997; Morin 2003; Herrmann 2013).

Thus, completing Step 3 gives the new employee the possibility of interpreting and reinterpreting the external environment, as well as directing a look within themselves to assess the internal environment, allowing them to prepare the fourth step of socialization coaching in which, possibly, we work with the construction of goals within a scenario known by the new employee.

7.3 Step 4: Reinterpretation of the New Employee Demands

Step 4 of the transformational coaching methodology refers to identifying interpretive gaps, in which the coachee is conducted to assess beliefs and habits, of which many are limiting (Wolk 2010). To the socialization coaching, the new employee is invited by the socialization coach to rethink his own opinions based on explanations in which he takes responsibility for his reality as the protagonist of his story. Once again, the questions that the coach addresses to the new employee will allow a reinterpretation of reality so that he knows that the world he sees is not always the world that others see.

In Fig. 8 it is reinforced that this is a moment of socialization coaching that must be thought of to modify the assessments that the new employee makes about the problem situations (Wolk 2010). The questions proposed in this step should be asked in a way that leads the new employee to question who is the protagonist about any of the situations that involve his life. The questions focus on the performance of the new employee in relation to the task, the group, or his role, allowing him to have

Transformational Coaching	Socialization Coaching: comments and adaptations
How could you explain what was worked, but from you, I mean, in the first person of the singular?	▪ *How could you explain the situation from yourself, in the first person of the singular?* ▪ *How could you explain the situation without involving others, taking yourself the responsibility for the situation?*
In other words, how could you transform the judgments and explanations of "victim" into explanations of "protagonist"?	▪ *What is your contribution to this situation? Who is responsible for you feeling how you feel?* ▪ *Is there another way to explain this situation?* The suggested questions set the tone for the moment and apply to Socialization Coaching, because they lead the new employee to question who has the active role on his condition; to confront and differentiate what is opinion and what is reality; to assess which individual beliefs limit him; to identify habits that are not productive. Finally, it is a time to reinterpret his situation and his demands and to confront his interpretation of reality with reality.
In what / how did you contribute to this situation?	It is suggested that the Socialization Coach appropriates the essence of the questions, adjusting them to the demands of the new collaborator, with the necessary subtlety so that they are not invasive or aggressive. However, it is up to the coach to explore and confront the opinions and judgments of the new employee in order to get him to assume the role of his trajectory in the socialization process. This protagonism applies to the task, the group, the role and performance, rescuing the individual characteristics of proactivity, creativity and innovation. The new employee is not a victim of the organization; he needs to take initiatives to seek socialization in order to be a protagonist within the organization.
What other way of explaining could you give?	For this reason, the proposed questions encourage the new collaborator to develop skills and competences related to the internal articulation, positively influencing his attitude towards the organizational environment. Questions have the power to provoke reflection on the new employee, developing the ability to establish verbal and non-verbal communication processes that contribute positively to socialization in an active and purposeful manner.

Fig. 8 Second Stage and Step 4 of Socialization Coaching

What wishes are you not honoring / what is your dream? What is your deepest interest behind this desire? What Truth are you hiding? When did that happen earlier in your life? How did you resolve or deal with the situation at that time? In what other circumstances does this happen or happened?	"There is no difficult coachee, there is an inflexible coach" is a common maxim in Coaching culture. And the questions suggested here will require the Socialization Coach's ability and sensitivity to propose questions that can be invasive, because they seek to explore what perhaps the new employee has not yet had the courage to reveal and, often, even to admit to himself. ▪ *What wishes are you not honoring / what is your dream?* ▪ *What is your deepest interest behind the desire?* These are questions that the Socialization Coach can ask to assess if there are signs of breach or violation of a psychological contract and that create in the new employee the desire for an early departure from the organization. The question seeks to extract from the new employee if the cost of being in the organization is too high for him or if the fact of being in the organization is aligned with desires and dreams. ▪ *What Truth are you hiding?* ▪ *When did this happen earlier in your life?* ▪ *How did you resolve or deal with the situation at that time?* These are powerful questions, but very personal. It is a time when trust between the new employee and the coach makes a difference so that the coach is allowed to do them. Questions about truths explore beliefs that can be limiting in the new employee. The coach must have the ability and the sensitivity to lead the new employee to articulate his / her beliefs in a way that they are not a limiting factor, but a potentializer. Questions about previous situations are intended to lead the new employee to find lived parallels that can be an example for him to proactively solve his organizational demand. All of this should be brought by the socialization coach to the organizational moment seen by the new employee, who will probably be linked to the task, the group, the role and the performance, always bearing in mind that he is part of other systems, such as family, friends and the community.

Fig. 8 (continued)

a perception of how he is at the moment; who he feels he is in that environment; how he uses skills to create in the organization; and what is his way of manifesting himself in front of others (Feldman 1976; Wilber 1997; Wolk 2010). In this way, when answering the questions, the new employee will be able to re-articulate beliefs (that, from limiting, become of expansion of the awareness and actions); and modify habits (that can contribute to make his walk in the organization a positive one) (O'Connor and Lages 2002).

The completion of Step 4 of the socialization coaching is an important moment within the entire process. It is understood that it is the completion of the exploration, understanding, interpretation, and reinterpretation of the demands of the new employee, associated with a deep approach to self-knowledge. The new employee assessed himself in the task, in the group, on the role and also in terms of performance, as well as situating himself in relation to himself, observing how he IS (as understanding), how he IS (as identity), what he CREATES, and how he MANIFESTS in this environment (Feldman 1976; Wilber 1997; Wolk 2010). The perception that the new employee is the one who determines the pace of evolution is rescued by assuming the protagonism of his organizational life.

The next step foresees the definition of actions, even though throughout the socialization coaching path, actions have already been taken, confirming the cyclical and non-linear nature in the socialization of the new employee, as well as in the coaching process.

8 Stage III: Expansion—Steps 5 And 6

The third stage of Wolk's (2010) transformational coaching describes Steps 5 and 6.

8.1 Step 5: Projection of Actions

Socialization coaching has already made the new employee assume his responsibilities in the socialization process, understanding and taking advantage of the welcoming and integration programs proposed by the organization in a proactive way (Herrmann 2013).

In Fig. 9, we want to demonstrate that this stage tends to be very valuable because the questions lead the new employee to design actions that can change the scenario. The rescue of what was previously worked will allow the new employee to start building actions aligned with the different dimensions of the systems of which he is part, such as the organization, family, friends, and the community with the individual pretensions (McLean 2012; Morin 2003). The problem situations will require the new employee to design concrete actions that positively modify the future.

The projection of actions that can lead the new employee on the path to becoming an organizational member is a reflection of the level at which he takes the lead to

<table>
<tr><th>Transformational Coaching</th><th>Socialization Coaching: comments and adaptations</th></tr>
<tr><td>What action alternatives do you see as possible?</td><td rowspan="4">It is the moment when Socialization Coaching tends to be more productive because it encourages the new employee to take action. As already said, this does not mean that this has not happened in the previous stages and steps, because they are all part of an integrated process of personal development. However, questions like, What action alternatives do you see as possible? What action strategies could help you get what you want? What are you going to do? charge from the new employee the creation of a future scenario and the planning of a present action.
The question How can you change what happened? serves as an alert for the new employee to continue to build on the present to change his future. And although the questions have been asked for the Transformational Coaching methodology, they are indicated for Socialization Coaching, since it also deals with a process of transforming an element external to the organization into an internal element of the organization. That is why the actions designed by the new employee will help him to actively engage in the process that is of interest to him, as well as to the organization and the working group.
Finally, the question What makes you think of them as alternatives? makes the new employee understand how each of his actions will impact the task, the work group, his role and performance, in addition to generating reflexes in family, friends and the community.
The questions in this step can be asked using the Goals at Work Grid as a basis and providing for the use of the Decision Ecology at Work tool which will be presented below .</td></tr>
<tr><td>What strategies / courses of action could help you get what you want?</td></tr>
<tr><td>What will you do?

How can you change what happened?</td></tr>
<tr><td>What makes you think of them as alternatives?</td></tr>
<tr><td>What would you do that for?</td><td rowspan="2">The Work Goals Grid, an exercise that is proposed at the end of Step 4, is a guiding element that reveals the goals of the new employee with respect to the task, the group, the role and performance. That is why carrying out the exercise is important for designing actions and taking responsibility for execution.
The question, What would you do that for? is indicated so that the new employee can identify the objectives behind the planned actions, without, however, sticking to judgments or</td></tr>
<tr><td>What do you expect as a result?</td></tr>
</table>

Fig. 9 Third Stage and Step 5 of Socialization Coaching

What do you feel as an impediment to acting? What do you see as an obstacle?	opinions. *What do you expect as a result?* leads the new employee to build possible scenarios with the results of his actions, involving the different processes foreseen in the socialization associated with the return he will have from them. When asking about obstacles, the intention is to alert the new employee to behaviors that may seem too assertive or, at times, even aggressive, producing results contrary to expectations. With the question, it is also possible to direct the new employee to take a position in relation to the practical or behavioral resources required to execute the proposal.
How can you tell your left-hand column wisely?	Finally, one last question in this block seeks once again to explore the hidden thoughts of the new employee that can offer new solutions on issues that bother him. Scrutinizing what is often not said so that it is said in a respectful and skillful manner tends to produce good results, including the goal that the new employee builds his way of becoming a full organizational member.

Fig. 9 (continued)

socialize as much as to be socialized. Socialization coaching is designed to give the new employee the proactivity and the protagonism about his socialization process in the new organization, avoiding the homogenization commented by Herrmann (2013), common with institutionalized socialization practices. And the present step of socialization coaching goes in the direction that the new employee shows what he is going to do to make this happen through his initiatives (Allen 2006).

Therefore, it makes sense to propose at this moment when proactivity, creativity, and possibly, innovation are revealed (Allen 2006; Herrmann 2013) arising from the behavior of the new employee, some support tool, such as role performance.

8.2 *Step 6: Role Performance*

In Wolk's transformational coaching methodology, (2010) Step 6 is intended for the use of simulation resources through hypothetical situations in the performance of roles. However, here it is up to the socialization coach to explore this tool or others that allow the new employee to take a deep look at his condition as an organizational member. The objective is for a new employee to be able to assess the impact of his actions on the performance indicators proposed by the HRM, as well as the impact on socialization processes in relation to the task, the group, the role, and a congruent assessment. The new employee will be able to create future scenarios to measure the impact on his level of job satisfaction, his organizational commitment, and the management of internal and external conflicts.

9 Stage IV: Closing—Step 7

In Wolk's (2010) transformational coaching methodology, the conclusion of the coaching process requires the coach's attention to consolidate learning. The last meeting is essential for the socialization coaching to end a cycle in which the new employee was willing to talk, think, expose, reflect, and propose actions that would help him to socialize in an active movement of becoming a full organizational member (Allen 2006; Herrmann 2013).

As defended, in each session, in each step, and in each stage, there was an opening, exploration, interpretation, expansion, and closure that also required care by the socialization coaching, as well as in the entire process (Wolk 2010). However, this stage composed of Step 7 coincides with the likely last session. Therefore, it is a moment of great responsibility of the socialization coach towards the new employee, who tend to be close to being a full member of the organization.

9.1 Step 7: Final Reflections and Closing

The seventh step of transformational coaching proposed by Wolk, (2010) suggests questions that lead the coachee to remember the path taken from the beginning, identifying the objectives set, the proposed goals, and the action commitments. The application suggested in this proposal for socialization coaching follows the same premise, helping the new employee to recapitulate the steps taken, the stages completed, and the commitments made with the aim of consolidating learning in order to contribute to make his socialization process faster and more effective. In this way, the last step of the socialization coaching process highlights the promises of action that the new employee made to himself, articulating the reflexes of his fulfillment in his surroundings (Fig. 10).

The completion of the process requires maturity on the part of the socialization coach to be clear that the path taken has contributed to the journey of the new employee in the organization. Here the questions posed throughout the process can and should be explored, looking for the connections between expectation, planned, and executed to perceive the repercussions in his organizational life. The questions and inquiries of the socialization coach should explore the new employee's ability to evolve within the organization towards socializing. Therefore, the new employee must be understood by the socialization coach as a complete, complex, and interdependent system that comprises and is integrated by other systems, among them the organization (Morin 2003; McLean 2012). Socialization coaching should contribute to this understanding on the way to becoming a full member. Thus, it is up to the

<table>
<tr><th>Transformational Coaching</th><th>Socialization Coaching: comments and adaptations</th></tr>
<tr><td>What did you learn?</td><td rowspan="5">The questions proposed by Wolk, (2010) suggested for this step need to consider the scope of Socialization Coaching, including the variables highlighted throughout the process. What did you learn ?, What do you think about the situation now? demand from the new employee a reflective process on the path taken, the situations faced, the demands presented and the intended objectives. The questions, What are you going to do with the situation? or What is your commitment to action with the situation? are questions that serve to consolidate the solutions thought up to date with the possibility of adding or deleting something. These are the situations with which the new employee lived in the period of three to four months, by which the Socialization Coaching was extended. It is a final invitation to learn and act, exercising the proactivity and creativity that can result in innovation in the face of the task, the group, the definition of the role and a congruent assessment of oneself. It is the end of the process in which the new employee is expected to have benefited from Socialization Coaching to increase job satisfaction, confirm his organizational commitment and have learned to manage his conflicts positively.</td></tr>
<tr><td>What would you do differently if something similar happened again?</td></tr>
<tr><td>What do you think now?

How do you feel now?</td></tr>
<tr><td>What will you do?</td></tr>
<tr><td>What is your commitment to action?</td></tr>
</table>

Fig. 10 Fourth stage and Step 7 of Socialization Coaching

coach and managers to recognize the new employee as a point outside the curve that needs to be received and welcomed in order to be part of an active movement without becoming homogenized.

Therefore, getting the new employee to build this relationship will enable him to reduce negative internal and external conflicts, exhibit the expected organizational commitment, and increase his job satisfaction (Feldman 1976; Wilber 1997; Morin 2003; Wolk 2010; McLean 2012). We are convinced that this alignment tends to reflect positively on the other indicators normally monitored by HRM, such as absenteeism rates, average promotion times, the quality of hiring, the minimum financial return required, the return from training and, especially index of voluntary or involuntary turnover in the socialization encounter phase (Allen 2006; Saks et al. 2007).

Finally, one last question can always be left for the self-reflection of the new employee: *is the organization better with his presence?*

10 We Are All a Point Out of the Curve: Recognize, Welcome, and Accept so that the New Collaborator Wants to Integrate Without Homogenizing

The new application of coaching as a personal development tool at a stage of the employees' organizational life in which it was not yet used, the socialization encounter phase, expands the resources available to managers and the HR area. It is the socialization coaching, an individualized and informal tactic regarding the context; and not sequential and variable in terms of content. As for the social aspects, it is an institutionalized investiture tactic and an individualized disjunctive tactic (Jones1986). The convergence of existing knowledge was explored in order to contribute to the solution of problems in the phase of the socialization encounter of new employees.

When implementing socialization coaching as another welcoming and integration tactic, it is expected that there will be:

- a decrease in turnover among new employees;
- the minimization of the feeling of insecurity generated by emotional factors of confrontation in the new environment;
- the decrease in the perception of role ambiguity;
- the redemption or prevention of breach or violation of psychological contracts;
- the increase in proactivity among new employees, leading them to assume the role of their socialization process;
- the increase in satisfaction in the work environment regarding the group and performance more quickly;
- the reduction of negative conflicts internal and external to the work environment;
- and the lack of homogenization of the individual characteristics of new employees, as defended by a wide range of authors, including Allen (2006); Cooper-thomas and Anderson (2006); Gruman et al. (2006); Ashforth et al. (2007); Korte, (2007); Rauber (2010); Herrmann (2013); Huhman (2014); Madlock and Chory (2014). Therefore, with the application of socialization coaching, it is believed that the new employee will have a faster and more effective socialization process, preserving the ability to create and innovate by encouraging his proactivity (Allen 2006; Herrmann 2013). Although organizations tend to align themselves with institutionalized and collective practices that are more strongly related to the decrease in the feeling of insecurity, as demonstrated by Ashforth et al. (2007), the adoption of individualized socialization tactics allows the new employee to have a more positive self-assessment by encouraging him to assume the protagonism of his choices on the way to socializing, according to authors such as Allen (2006) and Herrmann (2013). This is also a highlight of this work, which believes in the need to put people at the center of the socialization process, as well as the focus of organizational decisions. Welcoming and recognizing that each person is a complete, complex, and interdependent system, will allow the new employee to take the leading role in integration as a point outside the curve. There are no

people without organizations, so the question: why are people still not at the center of the actions of many organizations?

The theoretical foundation of coaching is based on the concept that an employee is a complete and complex system that integrates systems other than the organization. Therefore, it is up to the managers and the HR area to broaden the understanding about their collaborators, especially the new ones, analyzing them from a systemic perspective. What to do with the point outside the curve? Recognize them. There is nothing wrong with that. Recognizing and welcoming the other as a true other is what will lead the new employee to take an active path in his integration process, creating diverse, and competitive organizations. Therefore, the main result of the creation of socialization coaching are the benefits of its use.

After all, what is needed to welcome and integrate a new employee, leading him to be a full member of the organization in relation to the task, the group, the roles, and performance evaluations, maintaining high levels of job satisfaction and strong organizational commitment? There is no exact answer, however, to receive and recognize that each human being is a complete, complex, and interdependent system will help each new employee to take an active role in the quest to socialize with their full integration.

We are Humans. We are Multiple. We are unique.
We are all a Point Out of the Curve!

References

Allen, D. G. (2006). Do organizational socialization tactics influence newcomer embeddedness and turnover? *Journal of Management, 32*(2), 237–256. https://doi.org/10.1177/0149206305280103.

Anderson, M., Frankoverlgia, C., & Hernez-Broome, G. (2008). Creating coaching cultures : What businessences leaders expect and strategies to get there. In: *CCL Research White Paper*, pp 1–23.

Ashforth, B. E., Sluss, D. M., & Saks, A. M. (2007). Socialization tactics, proactive behavior, and newcomer learning: Integrating socialization models. *Journal of Vocational Behavior, 70*(3), 447–462. https://doi.org/10.1016/j.jvb.2007.02.001.

Barner, R., & Higgins, J. (2007). Understanding implicit models that guide the *coaching* process. *Journal of Management Development, 26*(2), 148–158. https://doi.org/10.1108/02621710710726053.

Boose, J. H. (1984). Personal construct theory and the tranfers of human expertise. In: *AAAI-84 Proceedings*, pp 25–33. Retrieved from http://www.aaai.org/Papers/AAAI/1984/AAAI84-030.pdf.

Cable, D. M., & Parsons, C. K. (2001). Socialization tactics and person-organization fit. *Personnel Psychology, 54,* 1–22.

Cooper-Thomas, H., & Anderson, N. (2002). Newcomer adjustment: The relationship between organizational socialization tactics, information acquisition and attitudes. *Journal of Occupational and Organizational Psychology, 75,* 423–437. https://doi.org/10.1348/096317902321119583.

Cooper-thomas, H. D., & Anderson, N. (2006). Organizational socialization: a new theoretical model and recommendations for future research and HRM practices in organizations. *Journal of Managerial Psychology, 21*(5), 492–516.

Cox, E., Bachkirova, T., & Clutterbuck, D. (2010). *The complete handbook of Coaching*. London: Sage Publications Ltd. https://doi.org/10.1016/0141-6359(80)90062-8.

Dass, P., & Parker, B. (1999). Strategies for managing human resource diversity: From resistance to learning. *Academy of Management Perspectives, 13*(2), 68–80. https://doi.org/10.5465/AME.1999.1899550.

Downey, M. (2010). *Coaching eficaz*. São Paulo: Cencage Learning.

Ellinger, A. D. (2004). The concept of self-directed learning and its implications for human resource development. *Advances in Developing Human Resources, 6*(2), 158–177. https://doi.org/10.1177/1523422304263327.

Feldman, D. C. (1976). A contingency theory of socialization. *Administrative Science Quarterly, 21*(3), 433–452. https://doi.org/10.2307/2391853.

Gallwey, T. (2000). *The inner game of work*. Toronto: Random House.

Gormley, H., & van Nieuwerburgh, C. (2014). Developing *coaching* cultures: A review of the literature. *Coaching: An International Journal of Theory, Research and Practice, 7*(2), 90–101. https://doi.org/10.1080/17521882.2014.915863.

Griffeth, R. W., & Hom, P. W. (2001). *Retaining valued employees*. Thousands Oaks-CA: SAGE Publications.

Griffin, A. E., Colella, A., & Goparaju, S. (2000). Newcomer and organizational socialization tactics: An interactionist perspective. *Human Resource Management Review, 10*(4), 453–474. https://doi.org/10.1016/S1053-4822(00)00036-X.

Gruman, J. A., Saks, A. M., & Zweig, D. I. (2006). Organizational socialization tactics and newcomer proactive behaviors: An integrative study. *Journal of Vocational Behavior, 69*(1), 90–104. https://doi.org/10.1016/j.jvb.2006.03.001.

Herrmann, A. F. (2013). Kierkegaard and indirect communication: Theorizing HRD, organizational socialization, and edification. *Human Resource Development Review, 12*(3), 345–363. https://doi.org/10.1177/1534484313482463.

Huhman, H. R. (2014). The 10 unique soft skills employers desire in new hires entrepreneuer. Retrieved March 1, 2017, from https://www.entrepreneur.com/article/234864.

Jones, G. R. (1986). Socialization tactics, self-efficacy, and newcomers' adjustments to organizations. *Academy of Management Journal, 29*(2), 262–279. https://doi.org/10.2307/256188.

Kim, T.-Y., Cable, D. M., & Kim, S.-P. (2005). Socialization tactics, employee proactivity, and person-organization fit. *The Journal of Applied Psychology, 90*(2), 232–241. https://doi.org/10.1037/0021-9010.90.2.232.

Knowles, M., Holton, E. F., & Swanson, R. A. (2005). *The adult learner: The definitive classic in adult education and human resource development* (Sixth). Boston: Elsevier. Retrieved from http://books.google.com.tr/books?id=J6qGsHBj7nQC.

Kolb, D. A. (2015). *Experiential learning: Experience as the source of learning and development (Second)*. New Jersey: Pearson Education Limited.

Korte, R. F. (2007). *The socialization of newcomers into organizations : Integrating learning and social exchange processes*. University of Minnesota.

Lane, D. A., & Corrie, S. (2007). *The modern scientist practitioner*. A guide to practice in psychology.

Madlock, P. E., & Chory, R. M. (2014). Socialization as a predictor of employee outcomes. *Communication Studies, 65*(1), 56–71. https://doi.org/10.1080/10510974.2013.811429.

McLean, P. (2012). *The completely revised handbook of coaching: a developmental approach*. San Francisco-CA: Jossey-Bass. https://doi.org/10.1007/s13398-014-0173-7.2.

Morin, E. (2003). *A cabeça bem feita: repensar a reforma, reformar o pensamento* (6th ed.). Rio de Janeiro: Bertrand Brasl.

Mosquera, P. (2007). Integração e Acolhimento. In A. Caetano & J. Vala (Eds.), *Gestão de Recursos Humanos: contextos, processos e práticas* (pp. 301–324). Lisboa: RH Editora.

Neale, S., Spencer-arnell, L., & Wilson, L. (2009). *Emotional intelligence*. London: Kogan Page.

O'Connor, J., & Lages, A. (2002). *Manual Curso de Certificação Internacional em Coaching*. Lisboa: Lambent.

Rauber, M. J. (2010). *As iniciativas de socialização e integração implementadas pela Universidade do Minho: O caso dos investigadores estrangeiros*. Universidade do Minho.

Robinson, S. L., & Rousseau, D. M. (1994). Violating the psychological contract: Not the exception but the norm. *Journal of Organizational Behavior, 15*, 245–259.
Saks, A. M., & Ashforth, B. E. (1997). Organizational socialization: Making sense of the past and present as a prologue for the future. *Journal of Vocational Behavior, 279*(51), 234–279.
Saks, A. M., Uggerslev, K. L., & Fassina, N. E. (2007). Socialization tactics and newcomer adjustment: A meta-analytic review and test of a model. *Journal of Vocational Behavior, 70*(3), 413–446. https://doi.org/10.1016/j.jvb.2006.12.004.
Sieler, A. (2010). Ontological coaching. In E. Cox, T. Bachkirova, & D. Clutterbuck (Eds.), *The complete handbook of Coaching* (pp. 107–119). London: Sage Publications Ltd.
Taormina, R. J., & Law, C. M. (2000). Approaches to preventing burnout: the effects of personal stress management and organizational socialization. *Journal of Nursing Management*, *8*(1), 89–99. Retrieved from http://web.a.ebscohost.com.
Van Maanen, J. (1978). Toward a theory of organizational socialization. *Research in Organizational Behavior*, *1*(1), 209–264.
Van Maanen, J., & Schein, E. H. (1979). Toward a theory of organizational socialization. *Research in Organizational Behavior, 1*(1), 209–264.
Whitmore, J. (2002). *Coaching for performance: Growing people, performance and purpose* (3rd ed.). London: Nicholas Brealey.
Whitmore, J. (2009). *Coaching for performance: Growing people, performance and purpose.* London: Nicholas Brealey.
Wilber, K. (1997). An integral theory of consciousness. *Journal of Consciousness Studies, 4*(1), 71–92.
Wilber, K. (2011). *Integral spirituality*. London: Shambala.
Wolk, L. (2010). *A arte de soprar brasas.* Rio de Janeiro: Qualitymark.

Coaching for All—New Approaches for Future Challenges

Carla Gomes da Costa and Andrea Fontes

Abstract Considering today's social challenges, the main goal of this chapter is to discuss what the role of Coaching has been in the organisational context in particular, and in the context of society in general. The discussion is mainly focused on the role of Coaching and how it has developed from organisational interests exclusively centred on Executive Coaching, to being a global instrument for individual psychological development.

1 Introduction

We are living in a time of great reflection due to the changes occurring to our planet and in our world, our society, our families and ourselves. We arrived in the twenty-first century to a world full of great innovation, but where uncertainty about the future is the rule, and where socio-economic differences are huge and widespread within each country and between countries. Robotics and artificial intelligence are hand-in-hand with the lack of regard and exploitation of children in some societies. The astounding evolution of technology and the consequent advances in human knowledge and competencies have far outpaced the moral values and laws currently in place to control them and prevent the misuse of certain scientific achievements and promote economic interests that could threaten the future of humankind. Indeed, achieving a balance between the use of different types of resources (including human) and respect for their innate nature has been a big challenge in several areas throughout the centuries.

The original version of this chapter was revised: Belated correction has been incorporated. The correction to this chapter is available at https://doi.org/10.1007/978-3-030-71105-4_7

C. G. da Costa (✉)
Instituto Superior Manuel Teixeira Gomes, Lisbon, Portugal
e-mail: carlagomescosta@ismat.pt

A. Fontes
ISCTE, Instituto Universitário de Lisboa, Lisbon, Portugal

C. Machado and J. P. Davim (eds.), *Coaching for Managers and Engineers*, Management and Industrial Engineering,
https://doi.org/10.1007/978-3-030-71105-4_4

The way modern societies have been organised after industrial revolutions has dictated today's socio-economic challenges. There are always two sides to any reality depending on the observer's perspective. The technological evolution has clearly reflected both sides—the side that brings improvements and has made life easier on this planet, and the other that is related to undesirable side effects, like the overuse of limited resources and the ensuing pollution. The pursuit of continuous economic growth has created a consumer society in which material goods and needs are developed as major assets to be perpetuated. This has led to the establishment of deep-seated values that influence the development of certain attitudes and behaviours that relate to a specific style of living. Nowadays, health problems brought about by stress and depression have called this into question and the pursuit now is for new ways of living and for new values that can orient new attitudes and behaviours that will promote people's well-being. The imbalance between work and personal life is the symptom of some mismatch between lifestyle choices and emotional health. Moreover, the ecological problems identified have also revealed the way natural resources are being overused, calling into question the very survival of human life on our planet.

In the light of all these challenges, what are the main skills and competencies that each of us needs to develop to be more able to survive in a world that is not only changing but that needs to change?

The 4th Revolution is already out there, with robotics and artificial intelligence altering work processes, silently transforming the way we work, and creating new jobs that will herald the end of others. Everyone needs to be aware that digital literacy is no longer merely a professional opportunity but has become a necessity for professional survival. Quite apart from honing technological skills and competencies, there is a need to develop greater psychological resources. This is where Coaching has a word or two to say since adequate psychological resources could help each of us to properly adapt to change and uncertainty in order to meet the big challenges we will face in the near future. The action-driven nature of Coaching, together with its solution-focused approach, combine to make an excellent tool for dealing with that changing environment (Grant 2014; Grant and Gerrard 2019). In fact, when focused on achieving goals and needing to overcome obstacles, the Coachee develops several psychological resources that will enable them to handle the particular situation that triggered the resource activation. What is more, they should then also be able to bring those psychological resources to bear in future situations in the form of self-efficacy, for instance.

2 Coaching in Society and the Relevance of Values

By the end of the twentieth century, Coaching had acquired a special status in the organisational domain. Coaching promoted the improvement of leadership competencies, which responded to the need to gain an advantage in an increasingly competitive organisational environment. Today, rapid change and a need to be creative in all areas of work are challenging Coaching to become an instrument that can help everyone enhance their personal and professional development (Cox et al. 2009; Theeboom et al. 2017).

In this global world context and specifically within the context in which we live, it is more than ever relevant to understand the role that Coaching needs to assume. The essence of Coaching can be conceived as a reflection process, whereby asking powerful questions can lead/challenge people to think differently than usual, in search of a pool of possible answers/solutions. With Coaching being a helpful relational instrument, it is relevant to understand the role Coaching can play today…in a pandemic crisis, like the one we are all currently living through. Indeed, what has the role of Coaching so far been in the big social crisis that we have been through? Given that Coaching has been enjoying a special status in organisations, how has Coaching served the various interests of different people, with different positions, functions, backgrounds, lifestyles?

The many, profound changes that have occurred in modern societies, especially from the 90s onwards, have affected organisations by creating specific circumstances that need to be adapted to. Having to deal with downsizings, restructurings and acquisitions, among others, has made organisational contexts much more demanding. Continuous change has made the world more uncertain and economically more competitive. This being so, executives are called upon more and more to improvise new competencies to creatively reorganise goals and develop further strategies and ways to motivate teams to respond to unexpected demands. All this contextualisation has opened the way for Coaching to proliferate as a new intervention that helps executives meet organisational demands and, indeed, even exceed expectations (For a review on executive Coaching, see Grant et al. 2010). The demands to achieve economic indicators quickly and consistently trigger a need to promote leadership skills based on psychological and behavioural skills (Grant 2014; Theeboom et al. 2017). In fact, knowing that Coaching is a customised intervention, it can be adapted to all types of contexts and cultures and, as such, overcome some of the barriers that might be found in traditional training. Additionally, if it promotes the development of psychological resources, it is easy to understand how the effects can be expanded beyond the immediate goals established in the Coaching sessions.

2.1 The Relevance of Values

With reflection being instrumental to Coaching, the first exercise will invite you to think about the relevance of values in your life and how they influence your decisions. Despite differences among individuals and between cultures, there are basic values that have been identified (Schwartz 2012; Borg et al. 2017). According to the social sciences, values are deemed to be social and individual organisers that motivate social trends, behaviours, and attitudes. For an individual, values function as basic principles (e.g. security; achievement; power) that guide perceptions of reality and judgement, which then lead to the subsequent action. Values have several characteristics such as their stability; their congruence or incongruence, among others, which reveal the various degrees of importance they have for a person or a particular society, and how they are directly implicated in decisions as they affect emotional inner balance. Values are related to what is really important to each of us in life, serving personal interests (e.g. achievement, hedonism) and/or social interests (e.g. benevolence, conformity).

Considering Schwartz's Value Theory (Schwartz 2012) the six features related to values are as follows:

1. Values are related to affect, meaning that people feel good or bad in accordance with a situation that may respect or disrespect their values.
2. Values have motivational power with regard to goal achievement.
3. Values are present in every area of an individual's life.
4. Values are the basic criteria that influence the perception and evaluation of every simple action, event, or person.
5. Values form an ordered system for people, revealing the importance of each value and underlying priority.
6. Multiple values could guide the same attitude or behaviour.

The following Table 1 sets out the ten broad values identified by this theory. The values identified are universal and based on human and social existence in accordance with the maintenance of biological organisms and group needs like survival and welfare.

Table 1 The ten values from Schwartz's value theory

Value	Needs served and goal intended
Self-Direction	Motivated by control and mastery needs, shows independent thought and action
Stimulation	Motivated by stimulation needs, shows a level of activation that looks for challenges, excitement, and novelty
Hedonism	Motivated by satisfying pleasure needs, looks for gratification
Achievement	Motivated by personal success and social approval demonstrates competent performance to stated social-cultural standards
Power	Motivated by prestige, control, or people dominance shows social esteem needs
Security	Motivated by needs for safety, harmony, and having stability in the individual and social sense
Conformity	Motivated by restraint of actions, avoiding social harm or norms of disrespect
Tradition	Motivated by respect and commitment to tradition, allows survival of society
Benevolence	Motivated by the affiliation of needs, improves others' welfare
Universalism	Motivated by the survival needs of individuals and groups, protecting the world and nature

Engaging in Coaching requires one to think deeply, to become aware of oneself, one's goals, one's life. This is what we are going to invite you to do now. Take some time, find a space, get a chair, a notebook and set out on a journey. Have you ever thought about what the main, deeply rooted values you hold are?

Based on Schwartz's Value Theory, please select and list according to your preferences the most important values for you, considering what is really important in your life.

List your values (at least 5)	*Give an example of how each one affects you in your life*	From 1 to 5 (1- the most important to 5- the least important)
1. 2. 3. 4. 5.		

Now that you have listed the most important values for you…. can you relate them to the way they influence or have influenced your life choices?

2.2 *The Relevance of Self-Awareness*

Given the greater challenges, life today presents, to achieve a better balance it is important to understand that human beings need to be aware that they functioning as a system. Each of us attaches varying degrees of importance to certain areas of our lives, in accordance with our values, goals and motivation. Understanding the dynamic of our own life system in order to ascertain precisely what these are may improve the balance between all the areas. This self-awareness can be considered one of the most relevant outcomes of Coaching (Bozer and Sarros 2012; Luthans and Peterson 2003) As stated by Theemboom and colleagues (2017), raising Coachees' awareness is one of the first steps in the Coaching process. In the beginning, it could be awareness of their feelings, thoughts and behaviours as well as the values that are the main pillars of the Coachees' motivation and that will help with the goal setting process (Grant 2014). At a more developed stage, this awareness would focus more on the competencies and resources that could be used towards the goal achievement.

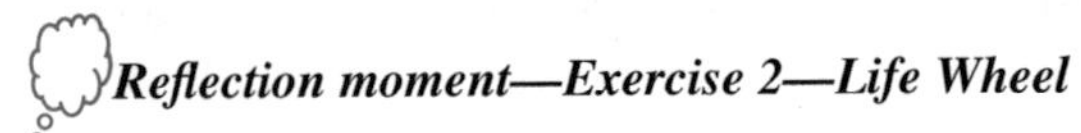

Reflection moment—Exercise 2—Life Wheel

- *Start by drawing a circle. Then, divide the circle into categories related to the main life areas that are important for you (or that you want to think about in more detail) such as: work, family, health, social life, leisure…*

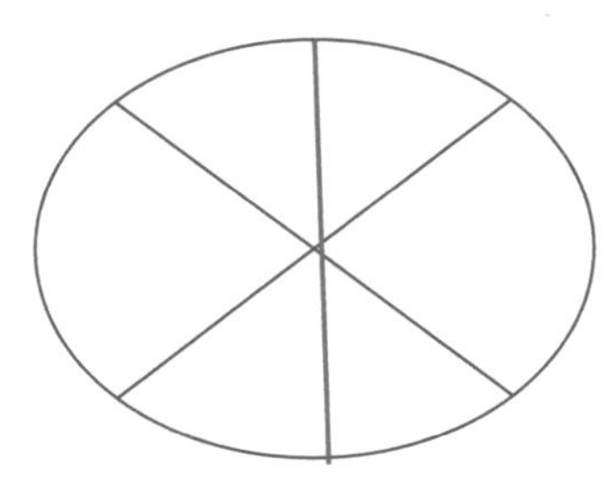

- *Having identified the most important categories, attribute the dimension that each has in your life (most to least important).*

- *For each area, identify 3 positive aspects and 3 aspects that you would like to improve*

- *How satisfied are you with each area (from 0% to 100%)*

- *Finally, identify the first area in which you'd like to make some changes from now on. What can you do immediately to improve your satisfaction with this area? Think about someone that could help you make this improvement. Are there any resources that you need to have/find? When do you want to start?*

____
____
____
____
____

Coaching in organisations has so far dealt with adapting to change but today, Coaching must provide answers to other needs. Living in such uncertain times in all areas of life prompts the need to promote Coaching at every organisational level. Everyone needs to be properly prepared and equipped to handle uncertainty, and by developing adequate psychological resources to help bring balance to people's lives they can be.

3 Goal Setting and Psychological Resources

Coaching, at its core, encourages the Coachee to be aware of and reflect on their resources and their strengths, as well as any possible obstacles and limitations they might face. The questioning process and the feedback provided by the Coach should expand the Coachee's knowledge about these factors and "force" them to both activate their resources in order to achieve the predefined goal and simultaneously find pathways to overcome any barriers. Several factors are relevant to this process. Here, we highlight two main factors: (1) the goal setting process that in itself implies the activation of a self-regulatory cycle and (2) the reflection and self-awareness that will lead to the finding of resources extremely relevant to goal attainment (Theeboom et al. 2017).

The goal setting process also drives motivation because when a goal is established, it determines a definite direction and stimulates the identification of the strategies required to attain it (Seijts and Latham 2005). In Coaching, the goals are set by the Coachee and shared with the Coach, albeit with a reinforced commitment (Fontes and Dello Russo 2020).

Personal resources come into play to achieve goals and stimulate personal growth and development (Gilbert et al. 2017; Ghielen et al. 2018; Xanthopoulou et al. 2009). We would like to highlight four specific personal resources: self-efficacy, hope, optimism and resilience which together comprise Psychological Capital (PsyCap; Luthans et al. 2007). Self-efficacy (the belief in one's domain-specific capabilities); hope (a motivational state characterised by agency and pathways); optimism (a positive attributional style about the future); and resilience (the ability to bounce back and effectively from adverse circumstances).

Previous research has shown that PsyCap can be malleable and reinforced through interventions (Luthans et al. 2006, 2010, 2008), as well as through individual Coaching (Fontes and Dello Russo 2020). In fact, the process of establishing a goal, finding and activating resources towards achieving it is the embodiment of hope. In other words, when the Coachee reflects on the necessary pathways to achieve the goal and manifests the agency to act, pursuing that goal is a manifestation of hope. In addition, and linked to the goal setting process, if the Coachee believes the goal can be achieved, he has positive attributions about the future, so optimism is present (Xanthopoulou et al. 2009). When the Coachee finds solutions to overcome the obstacles that inhibit them with regard to achieving their goal and develops strategies to better cope with change, they are, in fact, activating resilience (Youssef and Luthans 2007). And, finally, when the Coachee reflects on the critical situation they successfully overcame, the resources they brought into play and the solution they found, the Coachee is activating the self-efficacy that is a direct outcome, therefore, of the Coaching process (Bozer and Jones 2018). Within a self-regulatory cycle, learning from previous experiences helps build a new and constantly renewed resource package. Thus, through this continuous cycle of questioning and feedback from the Coach, together with the reflection and insight of the Coachee, psychological capital is activated and reinforced. (Fontes and Dello Russso 2020).

The following exercise, involving the four components of Psychological Capital, challenges the reader to experience this process. This exercise was adapted from the manuscript of Fontes and Dello Russo (2020).

Exercise 3—Psychological Capital

1. *Please identify a personally valuable goal (it can be one of those identified in exercise 2):*

____

(a) *Create concrete targets to measure success (can be time, value, number,…)*

____

____

(b) *Then identify sub-goals (if the goal defined above can be divided into smaller goals)*

____

____

____

____

2. *(OPTIMISM) Anticipate the best possible scenario regarding the goal achievement and be as detailed as possible:*

   Write what would happen

   ____

   ____

   what you would be feeling

   ____

   ____

   and thinking

   ____

   ____

3. *(SELF-EFFICACY) Recall and list positive/successful past events related to the goal, preferably where you have played an active role, but if you cannot recall any, they can be episodes from other people, that you observed.*

   ____

   ____

   What were the ingredients of that success?

   ____

   ____

   ____

4. *(HOPE) Generate multiple pathways to the identified goals, as many alternative pathways as possible, regardless of the practicality of implementation*

   ____

   ____

   ____

   considering the resources required to pursue each pathway, now rank them

   1st ____

   2nd ____

   3rd ____

   ____

5. *(RESILIENCE) Anticipate possible obstacles and separate what is under your control vs beyond your control*

_____________________ _____________________

_____________________ _____________________

_____________________ _____________________

Anticipate strategies and actions to avoid pathway blockage

____

____

References

Borg, I., Bardi, A., & Schwartz, S. H. (2017). Does the Value Circle Exist Within Persons or Only Across Persons? *Journal of Personality, 2*(85), 151–162.

Bozer, G., & Jones, R. J. (2018). Understanding the factors that determine workplace coaching effectiveness: a systematic literature review. *European Journal of Work and Organizational Psychology, 27,* 342–361. https://doi.org/10.1080/1359432X.2018.1446946.

Bozer, G., & Sarros, J. C. (2012). Examining the effectiveness of executive coaching on coaches' performance in the Israeli context. *International Journal of Evidence Based Coaching and Mentoring, 10,* 14–32.

Cox, E., Bachkirova, T., & Clutterbuck, D. (2009). *The complete handbook of coaching.* London: Sage.

Fontes, A. & Dello Russo, S. (2020). An experimental field study on the effects of coaching: The mediating role of psychological capital. *Applied Psychology,* published first online: 20 May 2020. https://doi.org/10.1111/apps.12260.

Ghielen, S. T. S., van Woerkom, M., & Meyers, M. C. (2018). Promoting positive outcomes through strengths interventions: A literature review. *The Journal of Positive Psychology*, 1–13. https://doi.org/10.1080/17439760.2017.1365164.

Gilbert, E. K., Foulk, T. A., & Bono, J. E. (2017). Building positive psychological resources. In C. L. Cooper & J. C. Quick (Eds.), *The handbook of stress and health: A guide to research and practice* (pp. 538–552). Hoboken, New Jersey: John Wiley & Sons. https://doi.org/10.1002/9781118993811.

Grant, A. M. (2014). The efficacy of executive coaching in times of organizational change. *Journal of Change Management, 14,* 258–280. https://doi.org/10.1080/14697017.2013.805159.

Grant, A. M., & Gerrard, B. (2019). Comparing problem-focused, solution-focused and combined problem-focused/solution-focused coaching approach: Solution focused coaching questions mitigate the negative impact of dysfunctional attitudes. *Coaching: An International Journal of Theory, Research and Practice*, 1–17.

Luthans, F., & Peterson, S. J. (2003). 360-degree feedback with systematic coaching: Empirical analysis suggests a winning combination. *Human Resource Management, 42*(3), 243–256.

Luthans, F., Avey, J. B., Avolio, B. J., Norman, S., & Combs, G. (2006). Psychological capital development: Toward a micro-intervention. *Journal of Organizational Behavior*, *27*, 387–393.

Luthans, F., Avolio, B. J., Avey, J. B., & Norman, S. M. (2007). Positive psychological capital: Measurement and relationship with performance and satisfaction. *Personnel Psychology, 60*, 541–572.

Luthans, F., Avey, J. B., & Patera, J. L. (2008). Experimental analysis of a web-based training intervention to develop positive psychological capital. *Academy of Management Learning & Education, 7*, 209–221.

Luthans, F., Avey, J. B., Avolio, B. J., & Peterson, S. J. (2010). The development and resulting performance impact of positive psychological capital. *Human Resource Development Quarterly, 21*(1), 41–67.

Schwartz, S. H. (2012). An overview of the Schwartz Theory of basic values. *Online Readings in Psychology and Culture, 2*(1). http://dx.doi.org/10.9707/2307-0919.1116.

Seijts, G. H., & Latham, G. P. (2005). Learning versus performance goals: When should each be used? *Academy of Management Executive, 19*, 124–131.

Theeboom, T., Van Vianen, A. E. M., & Beersma, B. (2017). A temporal map of coaching. *Frontiers in Psychology, 8*, 1352.https://doi.org/10.3389/fpsyg.2017.01352.

Xanthopoulou, D., Bakker, A. B., Demerouti, E., & Schaufeli, W. B. (2009). Work engagement and financial returns: A diary study on the role of job and personal resources. *Journal of Occupational and Organizational Psychology, 82*(1), 183–200.

Youssef, C. M., & Luthans, F. (2007). Positive organizational behavior in the workplace: The impact of hope, optimism, and resiliency. *Journal of Management, 33*, 774–800.

Prone to Follow, Eager to Lead: Millennials as the Ultimate Commodity on the Job Market

Diana Fernandes and Carolina Feliciana Machado

Abstract Reviewing literature, this study highlights the challenges organizations face once, for the first time, four generations coexist at the workplace—Veterans, Baby Boomers, Generation X and Millennials. Thus, it disserts around Millennials' construct focusing on their leadership and followership characteristics, discussing how these redefine organizational dynamic forces. Millennials' psychodynamics impel on management attention, foremost because many of these are based on the carry-over of their parents' mentality: we claim the "helicopter parenthood" experience (Twenge in Med Educ 43:398–405, 2009) developed into this group some aspects they transpose into the workplace, labelling them as the "Trophy Kids" generation (Alsop in The trophy kids group up: How the Millennial generation is shaping up the workplace. Jossey-Bass, San Francisco, 2008). As so, their early life experiences climaxed into a cohort that, we defend, is driven by a "pay-off" mindset always motivated by achieving, at the same time constantly needing guidance to averse risk and to feel validated by peers and (more importantly) superiors. This work suggests the relevance of implementing restructured organizations, directed at people and relationships, with a flatter hierarchical structure through upgraded flexible and cooperative manifestations of leadership, thus demanding advanced collective dynamics grounded on technologically learning and experimenting processes. Coaching, we argue, can be a good approach in regards to it.

1 Introduction

A *generation* can be defined as a subculture which reflects the prevalent values of a historical period, determined by significant cultural, political and economic

C. F. Machado (✉)
School of Economics and Management, University of Minho, Interdisciplinary Centre of Social Sciences (CICS.NOVA.UMinho), Braga, Portugal
e-mail: carolina@eeg.uminho.pt

D. Fernandes
School of Economics and Management, University of Minho, Braga, Portugal

C. Machado and J. P. Davim (eds.), *Coaching for Managers and Engineers*, Management and Industrial Engineering,
https://doi.org/10.1007/978-3-030-71105-4_5

developments, moulded by powerful external forces creating exclusive sets of values (Egri and Ralston 2004). Hence, members of a generational subculture rely on a set of shared beliefs, values, attitudes and logical processes, which provide the framework used by people within the group to think, act, reason, process information, socialize, work, organize and lead, instigating to societal changes via the idiosyncrasies of a given *generational identity* (Mannheim 1952; Ryder 1965).

For the first time in modern history, we have four generations (Veterans, Baby Boomers, Generation X and Millennials) represented in the workplace (Zemke et al. 2000), so, organizations need to make sure that, despite such differences, all members are working together towards the common goal of business' success (Kaifi et al. 2012). By understanding the characteristics and values of the Millennial generation, organizational leaders can adapt their leadership style to accommodate and engage them in the workplace as followers, and develop them into future leaders (Burkus 2010). Interestingly, organizations and researchers are just now starting to address such issues (Salahuddin 2010). Truth to be said: Millennials are here, prone to follow, eager to lead. Millennials are currently the ultimate commodity on the job market—whether an organization likes it or not, is ready or not, Millennial employees are roaring. The roar of Millennials is only getting glitzier, thus, innovative organizations must be prompt to reply accordingly.

We see the Millennial generation reshaping the world—the future won't just happen, it will be created, and primarily by Millennials (Balda and Mora 2011). In fact, as Generation X gave way to the leading population of Millennials in the workplace, a plethora of insights has been published to advise business leaders navigating such trend, advising on how to move beyond negative pre-existing stereotypes and an unappealing caricature for this generation (for a review, consult Stewart et al. 2017; Caraher 2015; Espinoza et al. 2010; Lancaster and Stillman 2010).

Aiming at adding crucial knowledge to the analysis of Millennials as leaders and followers in the workplace, this article, through multidisciplinary literature review (mainly based on insight from Psychology, Sociology, Anthropology, Economics and Management), disserts on Millennials' leadership and followership styles, providing solid insights into understanding this generation at the workplace. We provide coaching insights into both Millennial employees and their managers through which Millennials' skills and knowledge can boost organizational performance. These coaching insights also aim at improving the quality of Millennials' relationships in the workplace and their effects on organizational productivity, based as well on the fact that the challenges of a multigenerational workforce need to be addressed and managed the best way possible so that we can leverage both organizational and individual performances. Thus, this work provides a useful contribution aiming at retaining Millennials in the organization, developing them into strategic and valuable assets, its future managers and leaders. Therefore, this work provides an agenda for Millennial-focused workplace interaction research.

Our exposition articulates into five sections. First, we characterize Millennial generation interconnecting contributes from several scientific domains. Having this as background, in the second section, we present the main challenges this generation poses at the workplace, which we systematize as especially focusing on work–life balance and employee retention. Section three furthers the reflection interrelating the contributes of the two later sections by examining the leadership and followership patterns such generation presents at the workplace. Section four summarizes the insights so far provided by explaining them as grounding on the "helicopter parenthood" concept (Twenge 2009), and developing such argument we present the concept of "pay-off mindset" to characterize this generation, which we claim to be the basis to interpret their followership and leadership styles at the workplace and its respective challenges. Section five presents coaching insights, both for Millennials and for their managers, so that their performance in the workplace can be best managed and enhanced, aiming at retaining and developing such talent and best navigating the challenges posed by the multigenerational workforce we currently see in organizations. This coaching insights are oriented mostly at overcoming the main challenges in the workplace, as we have identified in section two, having as a rational how to best manage Millennials' features derived by the "helicopter parenthood" experience then crystalized into their "pay-off mindset", as exposed in sections one and five. Conclusions, limitations and further research challenges are, lastly, presented.

As Millennials will continue to enter the workforce until around 2022 (Hershatter and Epstein 2010), we can visibly state the relevance of this work. By understanding Millennials' leadership style, it may provide further contributes to design a work environment where leadership effectiveness can be maximized, fostering individual and organizational performance as it will potentialize a more accurate and efficient administration of the current multigenerational workforce challenges.

2 Mirror on Millennials—a Categorization

Millennials encompass a particular generational cohort, originally coined in the United States of America, born between 1980 and 1999, whose attributes, following Howe and Strauss (2000), defined a distinct subculture. There is a myriad of denominations for this group, the more frequently used are "Nexters" (Zemke et al. 2000), "Millennials", "Echo-Boomers" or "Thumb Generation" (Huntley 2006), "MySpace Generation" (Rosen 2007) and "NetGeners" (Tapscott 2009). Brooks (2001) as well suggests a very thought-provoking label for this cohort, the "meritocratic elite": Millennials' trust in institutions relies on an unbiassed system that ensures that accomplishment will be rewarded with validation and social mobility, progressing through an increasingly narrow tap.

As detailed by Howe and Strauss (2007), each generation has a *persona* with core traits—reviewing literature, we below identify the ones of Millennials:

- **Sheltered**: Millennials were deeply desired kids, raised by tremendously involved parents who coached them and interceded on their behalf, constantly protecting them in face of the adversities of life, instilling their importance to the family and the nation (Raines 2002). Therefore, literature exhibits a generalized idea that they were nurtured as a "special group", somehow the "golden elite", so that the enormous care and attention received from their parents turned them (overly) self-confident, empowered and optimistic (Cole et al. 2002). As a consequence, their high expectations in life are, sometimes, set as a pre-defined rigid plan, surrounded by family or peer pressure to accomplish (often, even to avenge) the life the previous generation did not have.

- **Confident**: Millennials are extraordinarily self-confident of their knowledge and capabilities (Harris-Boundy and Flatt 2010; Trzesniewski and Donnellan 2010). They are remarkedly trusting the future will be brighter, indicating increased levels of self-esteem and assertiveness compared to previous generations (Twenge and Campbell 2001). Indeed, as stated by Emeagwali (2011), this is the imperative they are driven towards, transversal to every aspect in their life, frequently promised since a young age both by their families and schools. Interesting to mention, Greenfield (1998) suggests this confidence has been lifted and sustained by an educational system with magnified grades and standardized tests, in which Millennials are superb.
- **Flexible**: Millennials see life in more flat terms, imprinting hope and cheerfulness on its experiences—they consider life as a path bursting with multiple prospects, all connected in a circular movement filled with unending opportunities via an ongoing experimentation and adjustment journey of continuous learning, testing and personal development. We can consider it as so based on the argument postulated by Brack and Kelly (2012), which recall this cohort's sensibility and tolerance once they are the most diverse generation to date (culturally, economically, politically, racially or ethnically).
- **Wired**: Millennials have grown up in the digital age, consequently showing greater familiarity than previous generations with communication devices, media and digital technologies (Gorman et al. 2004; Raines 2002). Given the rapid advancement of technology, victorious with the Globalization advent and boosted by the prevalent use of the internet, Millennials can quickly access, obtain and share information, which has decisively influenced their perspective on interpersonal relationships and workplace dynamics. Truth to be said, they frequently claim they, personally, could not live without the internet as an integral part of their existence (Cisco Corporation 2011). As a result, typically they are good at multitasking and willing to work wherever and whenever necessary: being more "wired" indeed provides them a competitive advantage turning Millennials into a valuable asset in today's technologically accelerated business environment.
- **Highly educated:** Millennials have been programmed to conceive education as a core domain to their success, based on the premise that in society there is a general assumption that having a degree will improve an individual's job prospects, statement imputed by their parents ever since a young age (Helyer and Lee 2012). Thus, they spend long years studying, being profoundly competitive and exigent with themselves, teachers and schools about their evaluation results. They consider educational life dynamics as an antechamber for work performance assessments. In fact, this generation is the most educated in history (Rikleen 2011), even though a trend in literature claims they possess insufficient decision-making and communication skills (Crumpacker and Crumpacker 2007).

- **Team Oriented**: Millennials are orientated towards a "collective ethos" (Hewlett et al. 2009), attributing vital relevance to hetero-judgment which is imprinted as a constant in their lives once they are driven to be inserted in groups and to develop through communal action. So, more than previous generations they value teamwork, wish collaboration, consensus and tight connections (Gursoy et al. 2013; Kaifi et al. 2012; Raines 2002), this being explained, in part, as a consequence of group-based learning since school years (Howe and Strauss 2007). Furthering this idea, empirical evidence reported by Pînzaru et al. (2016) supports the hypothesis that, compared to other generations, Millennials have better abilities to start/maintain relationships and to make a good first impression, as they excel the relevance of keeping social relationships – those are not necessarily face-to-face, as this generation growingly uses technology to interact: this reconfigures such interactions, because they are now more accelerated (as digital communication also is), allowing to contact whoever, from wherever and whenever. This is in line with Twenge's (2009) evidence that Millennials are more extravert, but also more anxious and frequently insecure. Also pertinent to remark in regards to this cohort's social skills is the fact that they have grown up in a diverse world and recognize the importance of learning and embracing new viewpoints to evade groupthink (Kaifi et al. 2012), but even if they are self-aware and confident on their own knowledge and capacities, they as well acknowledge the impact may be more effective by merging and coordinating ideas and efforts, thus being tolerant and open-minded.
- **Transforming**: If literature reports Millennials are alienated from fragmental themes such as politics (Tolbzie 2008), another trend defends they like to be kept informed and indeed take an active participation in all topics, freely providing their ideas and personal viewpoint or life experience (Hartman and McCambridge 2011). Consequently, they are often called "the loud generation", aiming at being noticeable and recognizable because they are driven to make a difference and wish to leave a footprint in history. It can be explained as this generation has been brought up into a world influenced by the Globalization, and it has been the promise made by their family and socialization institutions as they grew up: they would be the "golden generation", the "brighter future". Hence, Millennials are perceived as willing to thrive, creative in analysing and deconstructing the issue aiming at a better outcome, transforming their reality: as specified by Emeagwali (2011), they are often called "transformational" and/or "resilient". If, at a first glance, they can be seen as having less drive comparing to previous generations, caution shall be taken: they are driven by a wider and refined set of priorities. In regards to this, it is noteworthy to mention that, even during school life and/or work life (especially on its beginning), they engage in a plethora of social experiments (volunteering, gap years, among other initiatives), with which they may develop greater awareness of the world. So, we can call out for a certain hedonist trait in this generation, based on the argument that Millennials prefer collective action, wishing to be involved in projects that really matter and satisfy them, being civic-minded, hopeful and cheerful (Hewlett et al. 2009), also because during those

experiences they are likely to be exposed to cultural diversity and to become advocates for societal issues (Pew Research Center 2007).

- **Achieving**: Millennials are focused on accomplishing sophisticated expectations, thus commonly being called the "Trophy Generation" or "Trophy Kids" (Tolbzie 2008: 12) based on the emerging trend in sports and competition to reward. In an interesting perspective, this Millennials' characteristic can be explained as well by the fact that parents valued their children's opinions as consumers, involving them in adults' purchases (from cars to family vacations, for example), experiences which have contributed to Millennials' attainment orientation: they expect to find a job that is well paid and meaningful (Alsop 2008; Marston 2007) to be able to maintain the high life patterns they have been raised into, as they have depended on financial freedom and material goods provided by their parents but will require salaries to maintain such outlines (Pew Research Center 2007). Thus, Millennials' parents (mostly Boomers) are preparing (even pressuring) their children for financially rewarding career paths (Howe and Strauss 2003).
- **Pressured**: relating to the above, achieving is Millennials' *leitmotif*. They live in a competitive mindset, so have a very accelerated rhythm and tremendously high life expectations (Kaifi et al. 2012), but are often impatient, wishing to accomplish their goals quickly, continually hunting for communication to track their progress towards public praise, awards and/or bonuses. Although previous generations may have significant career ambitions, most Millennials strongly agree they are pressured to achieve just by the aim of achieving. As a result of intense socialization, Millennials place a high value on personal achievement, a constant throughout their lives and encompassing all its domains. Interesting in regards to this, Sadaghiani and Myers (2009) propose that intense and frequent socializing communication from parents about leadership also might emphasize this ongoing strand for personal achievement and material success in this generation, consequently not being surprising that, according to Pew Research Center (2007), 64% of Millennials report that getting rich is the crucial life goal for their generation.
- **Speedy**: Millennials have generally been raised in environments rich with feedback, praise and direction (Hershatter and Epstein 2010; Ng et al. 2010), mainly focussing on outcomes over processes (Thompson and Gregory 2012), being exasperated about becoming recognized as valuable contributors (Gursoy et al. 2008; Pew Research Center 2007). Thus, they prefer to move in a shorter time frame: an interesting line of explanation is provided by Dwyer (2009), suggesting that Millennials saw their parents being adversely affected by breakpoint situations in life, which then derived, for example, in divorce and/or corporate layoffs, resulting in a reluctance towards long-term vows, consequently demanding greater flexibility when it comes to professional careers. Indeed, Pînzaru et al. (2016) suggest this generation to be less focused on the processes and rules (often trying to overcome these), not appealed by projects/tasks involving details and monotony; hence, it is possible they report difficulties in delivering quality and on time because they might show a lower level of continual energy, low resistance to stress, a more fragile self-esteem and impatience when dealing with tasks requiring continued efforts.

- **Materialistic**: Protected by the ecommerce ecosphere, many Millennials have lived in times of relative economic expansion (Marston 2007) until the global recession began in 2008, subsequently, some authors sharply advocate this generation's frivolity and materialistic lifestyle claiming they so far did not feel life harshness to realize the overrated life standard they have been having. So, doubts arise around their capacity to deal with economic/financial adversity and whether aging throughout such difficulties will affect their expectations as they are just joining the workplace.
- **Roar**: Millennials are, hence, described as the "Look at Me" cohort, being overly self-absorbed, wishing power and gratification, projecting an unrealistic self-image (Pînzaru et al. 2016; Pew Research Center 2007). In fact, Twenge (2013) reports a dramatic intensification in the prevalence of narcissistic traits in this generation, compared to previous ones. This tendency can be observed right away in their propensity towards wide-spread dissemination of opinion, which, in line with Twenge (2009) and Twenge and Campbell (2009), may be consistent with a more ambitious and even egotistical group.

Condensing all the above, Raines (2002) postulates five motos to categorize this generation: "Be smart—you are special", "Leave no one behind", "Connect 24/7", "Achieve now" and "Serve your community".

3 Millennials in the Workforce—Challenges in the Workplace

Based on the above "X-Ray" of Millennials, it is clear that work organizations and management teams are growing awareness on the need to make changes to better accommodate this new work generation and capitalize on their talent (Stewart et al. 2017; Costanza et al. 2012; Emeagwali 2011; Hershatter and Epstein 2010; McGonagill and Pruyn 2010).

We advocate that Millennials address two main challenges in the workplace: one related to the *work–life balance*, other related to their *retention in the workforce*.

Work–life balance refers to the equilibrium employees establish in spending sufficient time at work while also allocating enough time on other tasks and pursuits, such as family, friends, leisure activities and hobbies (Smith 2010). Research has shown that creating a work–life balance promotes overall job satisfaction, increases productivity and potentializes employee retention (Gilley et al. 2015).

The past decades have been maturing a post-materialistic mindset in prosperous economies (Inglehart 2008), so that work stopped being valued in materialistic terms, rather being conceived as a broader experience. This is crystal when observed in Millennials' ideas and attitudes, as they consider more important to identify with the work community: in Goldthorpe et al.'s (1968) terms, to adopt a solidarity orientation towards work. In their view, the traditional workplace does not exist: they prefer horizontal communication and cooperative work based on technology and relationships' development, articulated in projects with clear objectives and room for creativity and innovation, instead of being bounded by a strict working schedule and fixed obsolete guidelines—they favour a workplace where they can entertain themselves while performing job tasks (Altizer 2010; Myers and Sadaghiani 2010; Twenge et al. 2010; Twenge 2009; Erickson 2008). So, and following the argument of Goldthorpe et al. (1968), this upgrade on work orientation springs from the individual's social and cultural background, which reconfigures personal values, being even more vivid for people who have more resources to invest in their leisure—Millennials crystalize this premise at best (Pyöriä et al. 2017).

Furthering these logics, empirical evidence reports Millennials are more family-oriented than previous generations (Hershatter and Epstein 2010; Cennamo and Gardner 2008; Carless and Wintle 2007; Martin 2005), so, career progression, family and leisure time are interconnected in a win–win–win game. Thus, there is evidence that these attitudes may act as a catalyst in organizations to change the traditional "workaholic" orientation and influence more human welfare workplace environments (Hewlett et al. 2009).

Millennials prefer to have a workspace at the office along with the option of working from home (Hewlett et al. 2009), therefore, Tulgan (2009) suggests that managers will be faced with the challenge to review how they perceive work time, which will require to cease paralleling presence with activity—nevertheless, the author also alerts that it is not unreasonable managers ask where Millennials will be working from and ensure the quality of their performance is not declining. Another important aspect managers shall keep in mind: before embracing long-distance work relationships, organizations shall carefully examine how to ensure liability and how to address this workers' need for frequent face-to-face interaction (Hershatter and Epstein 2010). This climaxes in the premise that Millennials believe that having the freedom to choose when and where they work is vital (Tulgan 2009) because a source of dissatisfaction at work amongst these employees consists precisely on the conflict between their work–life balance expectations and how they perceive to be supported by management teams in regards to it (Gilley et al. 2015): empirical evidence reports that few managers respond adequately.

To fulfil this demand, organizations can offer flexible work arrangements (variable start and end times, telecommuting, working from home, part-time hours and special summer hours or vacation days) rather than exclusively focus on salary. Technology can be a prodigious ally because it frees employees to work at a time and place convenient to them, having the additional advantage of being an environmentally friendly approach, consistent with Millennials' affinity for pressure towards societal awareness and planet protection. This way, organizations would achieve a mutually supportive situation by which they leverage internal branding as employee-friendly and benefit from more healthy and motivated workers, which will provide lower turnover rates and healthcare costs (Smith 2010). Gilley et al. (2015) provide a very thought-provoking contribute: based on the premise that work–life balance does not end on Human Resource Management practices such as the above enumerated, the authors point out that through other processes (namely, coaching and mentoring) managers would as well be able to enhance higher levels of productivity, lower rates of absenteeism and enhanced employee retention, subsequently, organizations would be allocating efforts on a more holistic approach towards work–life balance and personal development.

Job satisfaction is a virtuous proxy for *employee retention* (Coomber and Barriball 2007; Tourangeau and Cranley 2006). As a concept, it grounds on the dyad consubstantiated by an evaluative (cognitive) and emotional (affective) reaction to work environment and tasks (Hodson 2004; Hulin and Judge 2003), so it assumes that employees evaluate all aspects of their job outlook, consider alternatives and through an internal calculus arrive at an overall evaluation of their jobs' quality.

For Millennials, it is crucial that the dyads person–job and person–organization fit to create meaningful work-related duties (Nolan 2015; Scroggins 2008), as this generation is decisively marked by a trade-off mentality as it is clearly imprinted in their minds the fact that they have invested a lot of money and time in completing their studies, being displeased if hired for lower salaries and responsibility in face of what they expect. Consequently, Millennials seek employability and flexibility rather than job security, they call out for opportunities to do something substantial and exciting that helps others or makes a difference—hence, they will apply to a position/join an organization solely because they want to (Brack and Kelly 2012; Tulgan 2009; Alsop 2008; Oppel 2007). It is then not surprising that Millennials are mainly focused on an individual lifestyle aim, so, they privilege opportunities for personal development rather than lifelong employment, being less committed to one single employer comparing to previous generations (Marston 2009, Broadbridge et al. 2007). Consequently, Stewart et al. (2017) empirically report the incongruence between Millennials work-related behaviour and objective organizational commitment, measured in terms of turnover rates, reinforcing the need to address problems in Millennials' tenure. Tulgan (2009) affirms that leaders must prove how working in the organization helps to make an impact, evidencing not just how the organization itself is affecting the world, but as well stressing how Millennials' job is part of that effect. Hence, Human Resource professionals must integrate internal branding in the recruitment and selection process by attracting and developing targeted individuals, articulating their personal values with the branding strategy and performance management through continuous coaching and feedback (Nolan 2015, Özçelik 2015).

Surprisingly, empirical evidence found by Pyöriä et al. (2017) provides no confirmation that this generation is willing to change jobs, even though they are better placed than before to make independent choices and to get employers' ferocious competition for their services: on the contrary, the results seem to indicate that Millennials are highly committed to the workplace, but only once they have found their own arena.

4 Leadership and Followership Patterns in Millennials

Leadership refers to the set of harmonized actions designed and implemented by the individual who carries the primary responsibility for performing these in the group and who is given the mission of coordinating such activities (Fiedler 1967), thus guiding followers towards common goals (Bryman 1992; Pfeffer and Salancik 1975). Being leadership defined in several ways, leadership styles have as well multiplicated because they have developed grounding on several dimensions (Ismail and Ford 2010). Nonetheless, followership approaches have been studied less extensively than leadership ones, often because they usually associate with negative connotations (Bjugstad et al. 2006), even though leadership approaches would be inexistent without followers, as such relationship grounds on a dyad (Burns 1978; Meindl 1995; Howell and Costley 2001; Mushonga and Torrance 2008).

Since the negotiation is initiated by the leader and reciprocated by the follower, a leader's effectiveness is significantly influenced by the followers' permission (Avolio et al. 2009; Graen and Scandura 1987; DePree 1989; Graen and Uhl-Bien 1995), based on the leader–member exchange (LMX) theory (Dansereau et al. 1975). So, organizations should strive to match their leader's approaches with their followers' leadership preferences (Chou 2012).

As a major topic in this discussion, it is central to highlight that communication is a major theme for Millennials. Not only Millennials expect constant communication with supervisors (yet autonomy in carrying out their responsibilities), they in fact want communication to be a dialogue, different from what previous generations projected (Balda and Mora 2011; Hole et al. 2010; Myers and Sadaghiani 2010; Gursoy et al. 2008). These employees tend to be particularly confident on their skills (Harris-Boundy and Flatt 2010; Twenge and Campbell 2001), perceiving their predecessors, managers included, as mentors rather than content experts. Therefore, a delegating leadership may boost their accountability and personal fulfilment on the job, then transfiguring into leadership effectiveness, leveraging employee retention and organizational performance (Bjugstad et al. 2006).

Consequently, Millennials value leaders that can assume as mentors and/or that will connect them with mentors they can trust and progress from (Dulin 2008), actively listening to them, valuing their opinions and incorporating their own needs and viewpoints in the decision-making process. As this generation is networked, constantly shares knowledge and prioritizes personal bonds, according to Balda and Mora (2011), it is relevant to notice that Millennials spend considerable attention and time within organizations positioning their relationships to maximum advantage, not being intimidated by seniority or status. In fact, interpersonal communication is one of the main areas of difference between Millennials and other generations, as they evidence an entirely different perspective about privacy (Credo et al. 2016)—for example, regardless of their low-level positions, these employees feel a need to be kept in the loop of information. Balda and Mora (2011) thus defend that the impatience, intense scrutiny and an apparent impudence in requesting a voice found in Millennials may be a symptom of their voracious and relentless drive for tangible impact, their wish to crystallize and objectivate every action they take.

Millennials also prefer projects/groups/organizations with central decision-making, clearly defined responsibilities and formalized procedures. This, according to Hershatter and Epstein (2010) may find its rationale on Millennials' socialization patterns, specifically at school life, which was marked by a system plainly circumscribed, grounded on periodic evaluation scales with rakings in which these generations could easily and quickly situate to have an overview on how it was performing and to which aims could aspire to, having specific guidelines, tools and procedures (exams, group works, classes, assignments, etc.) to achieve the desired goals. Nonetheless, Millennials are motivated by freedom in the workplace and may be more loyal to leaders and organizations if provided, stimulated and recognized such freedom. Myers and Sadaghiani (2010), Hewlett et al. (2009), Alsop (2008), Gursoy et al. (2008), Mushonga and Torrance (2008) and Bjugstad et al. (2006) recognize such aspect as the ground base for this generation's personification of an exemplary followership style in the workplace, as they by their own impetus work well cooperatively and are constantly available to those who interact with them, engaging in their leaders' decisions not being hindered by their position or/and experience. Hence, Raines (2002) suggests the "Let's have fun!" moto to root the communication approach aimed at this goal, explaining that a bit of silliness, a little humour, even a little irreverence in the communication style from leaders towards their Millennials followers will make the work environment more attractive to them, once this generation is marked by the digital era and the triumph of mass media.

We can notice a trend in the literature defining this generation by manifesting narcissistic traits. As so, it is relevant to highlight that followership patterns of Millennials do not deeply identify with ideologies, concepts or structures, instead, they are constructed around personalities as their reference points, admiring idols (Stewart et al. 2017). From a follower perspective, Balda and Mora (2011) suggest that Millennials may attach to leaders whom they respect not only because they have formal power but also because they spot in such people appreciated personality traits and life experiences, which they then project into, wishing to imitate (Hershatter and Epstein 2010; Ng et al. 2010; Lancaster and Stillman 2010). It is the manager that supports work–life balance, impacts on performance appraisal and carries positive internal branding through daily operations: therefore, such employees' retention lies within the manager and the respective relationship, as stated by Thompson and Gregory (2012). Gilley et al. (2015), Nolan (2015) and Thompson and Gregory (2012) then suggest customizing leadership: managers shall be coaches providing encouragement and guidance to their team members. Nonetheless, a large part of the problem is that frequently immediate supervisors are individuals who excel in the job's technical aspects and have several years of experience, however, also often lack managerial skills, training and experience in dealing with the demands of multigenerational employees (Thompson and Gregory 2012; Timo and Davidson 2005).

On the other hand, Millennials are inevitably beginning to play an important leadership role in workplaces, which poses unique challenges to organizations (Smith and Nichols 2015).

According to Chou (2012), Millennials have been shown to exhibit an inclusive and participative leadership style where, as core, roots the value of instant feedback: literature states that Millennial leaders ground on a win–win communication approach, as they gather and share information eagerly through recurrent, positive, constructive, equitable and fluid communication flows (Gursoy et al. 2008; Howe and Strauss 2007; Marston 2007; Martin 2005; Zemke et al. 2000; Tapscott 1998), as this dynamics may foster the team member's inclusion in decision-making, actively asking for their suggestions and openly discussing organizational issues with them (Chou 2012; Chen and Tjosvold 2006).

Gibson et al. (2010) examined the differences in management values between generations concluding that there were more matches than mismatches, nonetheless contending that the top five values orienting Millennial leaders were family security, health, freedom, self-respect and true friendship. Howe and Strauss (2000) seem to share these ideas, as they advocate that Millennials possess the ability to perform with effectiveness in leadership roles, reflecting some of the characteristics earlier generations' leaders displayed. On another line of analysis, empirical evidence collected by Hartman and McCambridge (2011), focused on university students from Millennial generation, reported that although Millennials have been characterized as technologically erudite and multitasking, they are deficient in interpersonal communication skills (oral and/or written). On this line, evidence collected by Underwood (2007) synthesizes that Millennials are hardworking, visionary and socially focused, however, are susceptible to (unreal) expectations that impel on them an overrated sense of entitlement, which may render them into ineffective leaders. Indeed, Sadaghiani and Myers (2009) proposed that socializing communication from parents about leadership, often provided through self-centred discussions focusing on extrinsic benefits associated with leadership, might have emphasized personal achievement and material success stimulating Millennials to lead for selfish reasons rather than for the intrinsic satisfaction of self-development, by a desire to benefit followers and to satisfy the organizational common good.

5 The Helicopter Parenthood, Core of Millennials' Followership and Leadership Patterns

Millennials had parents who were more available to them than the previous generation, thus, they expect more supervision, feedback, clear goals, structure and mentoring, as well as a focus on their individual development in the context of work (Thompson and Gregory 2012). In fact, managers essentially fill the role of guiding parents as soon as Millennials join the workforce (Hershatter and Epstein 2010; Ng et al. 2010). Thus, we suggest that the "helicopter parenthood" (Twenge 2009) which decisively marks Millennials childhood is crucial to understand their dynamics at work, their followership pattern and leadership style, grounding several studies which portray this generation in rather paradoxical images. But this is a complex concept and literature exposes well two trends in regards to the parental effect on Millennials socialization: while some analyses contend that Millennials are self-centred (Twenge 2009), others argue they value community, civic duty, collaboration and volunteerism, having been bombarded with messages that they should serve their community ever since a young age (Greenberg and Weber 2008; Jacobson 2007; Raines 2002). As so, there is concern that for some of the most educated and high achievers the socialization of parents may drive them to strict and piercing levels of achievement, once Generation Y members have had parents who educated them to believe they are better than others, controlling many aspects of their children's lives, raising them to be competitive. As so, it could be argued that those aspects possibly limited their altruistic side, even though (and interestingly) Millennials' life experiences as young people, particularly in their volunteerism, may prevent this self-centred egoism from driving their future conduct as leaders (Sadaghiani and Myers 2009).

The construct of a leadership style for Millennials towards altruism and common good may be explained if taken into consideration that Millennials are the heirs of Globalization, having been born into that new global order. Thus, in what is becoming a much smaller world, they also need to learn how to work with others from diverse backgrounds, and successful leaders will need to have the skills to build and be a part of communities they have not imagined so far. Having always lived in this globalized world, Millennials display a great desire to be part of something bigger that will bring change, indeed they have a strident awareness that actions shall be taken to address the issues currently facing the world, aiming at turning the present (but foremost the future) positive. They want to have more meaning in their lives and demand to connect with the purpose and mission of any organization they take part in. Nevertheless, as Millennials are speedy, it may as well be argued that these think change happens much more easily than in old generations, so Millennials are often disappointed in prolonged efforts and long-term plans.

Another interesting point to explain this altruistic outline of Millennials in their followership and leadership patterns lands on the self-help literature and the growing focus on self-analysis, a cultural trend that started in the 1980s and continued until this day. These inputs, often brought to their discovery by parents, influenced particularly the way these youngsters behave. However, the effects of helicopter parenting and of the focus on self-discovery lead them not only to build high levels of self-trust but also to be dependent on others (Lythcott-Haims 2015). For Millennials, this increased self-esteem and assertiveness supports their meritocratic mindset, their belief in the right of individuals to succeed and contribute in the workplace regardless of their background (such as years of experience on the job) and to treat people as individuals rather than members of groups. However, this may also contribute to their impatience and lack of perseverance (Campione 2015), as it boosts their need for constant and immediate recognition. Combined with the finding that Millennials are also risk-averse and have difficulty dealing with ambiguity, they often require a clear path to success with visibly defined steps and expectations, as well as constant feedback from their supervisors (Myers and Sadaghiani 2010; Gursoy et al. 2008).

Another contribute in regards to the influence of Millennials' "helicopter parenthood" (Twenge 2009) on their followership and leadership styles can be found in Erickson (2008), as the author reports interesting relationship building between Millennials and Baby Boomers. Beekman (2011) and Alsop (2008) agree with that strategy of implementing cross-generational mentoring schemes in organizations, encouraging to, if possible, allow Millennials and Boomers to seek each other out freely and on a constant basis so that they can share knowledge and experiences. Millennials are the first population to have been fully immersed in mentoring programmes throughout their lives, even starting such programmes in elementary school. Furthermore, Erickson (2008) explains, such relationship may recall Millennials of their own familiar experience: as very nurtured children, they may see in Boomers a reference in work as their parents are in their homes, reaching out to them as to have a safeguard, gathering guidance and counselling to learn and to be able to overcome obstacles, thus, successfully achieving the desired results. Stressing this idea, Erickson (2008) continues to state that Boomers indeed appreciate being sought after for their knowledge and life experience, being more inclined to help because they frequently have Millennial-aged children themselves. However, this can create tension with managers, who occupy the middle hierarchies that Millennials sometimes hurdle over, so, Millennials' direct managers should be kept informed on and involved in the relationship (Tulgan 2009).

Summarizing the above described, and as seen in the first section, Millennials were raised under heavy supervision: this generation didn't grow up in a world where kids left the house on their bikes in the morning, met their friends and played freely in the park, returning in the evening in time for dinner. They were raised by devoted parents, who often infatuated their life desires into their children, so that their words, actions and emotions were given a large amount of respect by doting parents. As children and adolescents, their opinions were perceived as valid and contributed to discussions with parental figures (Alsop 2008). They were faced with constant academic and free-time activities, carefully planned and supervised by their parents, aiming at fully developing their potential both in technical as well as in social skills, at the fullest possible (Brack and Kelly 2012).

Thus, they have tremendously high life standards, expectations to be fulfilled in order to provide them a comfortable lifestyle, often the one their parents so much wished but unfortunately did not have. So, they want to be rewarded by their utmost dedication in every single minute spent through their lives in every single action they engage into, as they are *compulsive achievers*, driven by a highly competitive mindset, as they were raised in a clear child-centric era (Tulgan 2009). Hence, we suggest that the dynamics of the "helicopter parenthood" (Twenge 2009) decisively impacted Millennials' behaviours, ideas and expectations towards work and the workplace, developing a generating of compulsive achievers, the "Trophy Kids" in Alsop (2008) words. We agree and expand Alsop's (2008) labelling suggesting that this generation, due to the "helicopter parenthood" dynamics, can indeed be systematized through the concept off "pay-off generation", as Millennials are driven towards endless achievement, raised to perform a plethora of activities and to excel in a myriad of skills, thus growing up exposed to a socialization process impelling into them a continuous competitive attitude, constantly driven by a "trade-off" approach, conceptualizing life through an utilitarian lens, always seeking to maximize their outcomes and to expand their Production Possibility Frontier. Therefore, Millennials are constantly wishing to see their efforts rewarded and recognized the fastest and fully possible, so, they constantly aim at paying-off not only their personal efforts in direction to the "brighter future promise", nurtured by their "helicopter parents" all along with their life, but (and consequently) as well to pay-off their parents' sacrifices in order to provide them background so that they could profit from such experiences and thrive towards the desired outcomes. This gives also room for discussion, as we have seen in some trends in the literature, to characterize this generation as having an infatuated self-awareness, often conceiving themselves as a "meritocratic elite", hence reporting narcissistic traits.

6 A Multigenerational Workforce: Coaching Inputs to Human Resource Management

To be efficient in today's world, organizations must be capable of managing a multigenerational workforce with diverse and fluctuating beliefs, work ethics, lifestyles, values, attitudes and expectations (Barzilai-Nahon and Mason 2010; Niemiec 2000). There are indeed concerns among researchers that the characteristics of the Millennial generation will have an impact on employee and organizational performance (Alsop et al. 2009; Alsop 2008; McGuire et al. 2007). Nevertheless, empirical evidence is blurred in regards to Millennials' impact on the organizations' multigenerational workforce. There is a trend in the literature claiming that Millennials' characteristics may complicate, and potentially disrupt, workplace interactions with employees of other generations, hence negatively impacting coworkers and organizational processes, not supporting the view that recent economic conditions may push these generations to be more compliant than previously speculated they would be (Alsop et al. 2009; McGuire et al. 2007). Nevertheless, Pyöriä et al. (2017) state that the work orientation of the various generations composing the workforce shows more signs of continuity than of conflict, not supporting the claim that Millennials will be forcing organizations into radical changes, based on the *generational contract* of our society, as per which the common denominator in the continuum of generations is reciprocity.

Understanding Millennials will help managers to engage them as followers and eventually develop as leaders (Chou 2012; Myers and Sadaghiani 2010). In order to attract, develop and retain Millennial talent, organizations will be faced with the challenge of adapting their structures, practices and procedures to accommodate Millennials, which involves negotiating a complex set of dynamics for multigenerational participants, directing the emphasis on two critical factors: *purpose* and *action* (Balda and Mora 2011).

Millennials were raised with constant direction and feedback, so expect it to continue in the workplace: as so, we propose that coaching can be a useful tool to develop talent, with potential for positive impacts in what refers not only to keep Millennials engaged in their work but as well to best manage the interconnectedness of challenges posed by a multigenerational workforce that nowadays evidently marks the workplace. Coaching aims at helping professionals to have clarity and evaluate their lives progression path, acknowledging them about their objectives, values, dreams and desires, by means of presenting new options and guiding them towards change, via a dynamic, collaborative and progressive dialogue setting an ongoing process in which constant feedback is provided. Managers can indeed implement coaching and/or mentoring approaches to encourage cross-generational relationships—generational differences are, then, narrowed and innovative viewpoints shared by both parties, developing Millennials' leadership and followership skills, boosting a more solid and homogeneous workforce (Murphy 2012; Spiegal 2011). Based on this argument, and foremost based on Millennials traits and their impact on organizational dynamics, we below dissect the main multigenerational organizational challenges we suggest the implementation of coaching approaches shall take into account.

(a) **Recycle Learning and Training**

To accommodate these new workers, there is an urgent need to redefine academic curriculum, socialization experiences and ground values, work to be done in the first instance within the family/peer group, as well as at the classroom (Howe and Strauss 2007) Nonetheless, organizations shall also be aware and willing to review their learning and trainings schemes, as they are very important structures in an individual's life. This is crucial foremost taking into consideration this generation's traits, right away from their digital mindset, which turns them more wired and sensible to visual *stimuli*, quick learners, being curious and searching for information in every topic of their interest, to have a more grounded knowledge of several subjects and providing their opinion in a more visible and loud tone, this acting as a vehicle so that they can actively take steps into this world's change towards progress. Howe and Strauss (2007) indeed claim that this presents a clear evidence on the need to foster an organizational learning environment aligned with Millennials seemingly preferred learning style, which we can label as being a more "hands-on" approach.

(b) **Kings and Queens of the Digital Realm**

Digital technologies are indigenous to Millennials, as so, they are often labelled as "digital natives" (Prensky 2001): this generation has a different way of thinking and processing information because they have been raised fluently embedded into the language of computers, video games, information management and sharing, networks and the Internet, growing up absorbed by technology habitually without authority figures controlling their access to information (Espinoza et al. 2010).

Tapscott (2009) believes that the Internet acts as a globalizing force that flattens the world, makes distances shorter and eliminates distinct local characteristics youngsters might have had in other times, idea supported by Pînzaru et al. (2016) and by Palfrey and Gasser (2008). Hence, while technological developments have played a primary role in shaping the ways Millennials learn, interact and communicate, changes in the ways societal institutions think about and treat this generation have had, as well and for sure, a no less weighty impact on Millennial mindset.

Hershatter and Epstein (2010) alert, as a consequence, that when a quick answer is eagerly available, Millennials may not be motivated to search deeper in the problem's root, also being less prone to take time understanding others' perspectives. Millennials' social interactions are marked by social networking and participatory culture, which have had a pervasive impact on the way such generation interprets the world and constructs identity. Small and Vorgan (2008) highlight significant disparities in brain activity among generations, cataloging it as the "brain gap": Millennials are more effective in some arenas, like multitasking, filtering information and responding to visual stimulation, but less proficient in face-to-face interaction and deciphering nonverbal communication, also having despised the capacity to engage in deep and sequential focusing (Tapscott 2009).

The new processes by which knowledge is constructed through the Web and the use of communication devices have heavily influenced Millennials' course of knowledge search, assimilation and sharing. This open and plane access to knowledge acquisition, transfer and cocreation, the process also known as "social learning", is evident and typical of the networked environment in which Millennials have been raised, which leads to the advent of organizational perspectives focussed on *collaborative communities* (Prusak 2011). This is so as it challenges organizational structures, management tasks and responsibilities, the works' essence and the workforce characteristics, and such collaborative organizations provide the environment for sustained innovation and creativity along with constant knowledge creation and transfer, boosting organic coordination and connectivity, providing a powerful source of competitive advantage. These viewpoints about generational competencies shed light on an interesting dilemma: in fact, the current business arena often requires imminence, valuing the ability to efficiently retrieve and communicate in a concise, simplified and assertive mode. Yet, the organizations' complexity (and as well the intricated interconnectedness of the environments in which they operate) demands a more nuanced, deep and informed framework for investigation. Following Hershatter and Epstein (2010), if Millennials are going to become valued knowledge workers, they must learn not only which information to gather, create, use and share but also how to examine and understand it in context.

(c) The Teamwork Fallacy

Millennials may find excessive comfort in team-based direction and decision-making, thus avoiding risk associated with independent thinking and decision. Let us here recall that this generation is used to be guided by dotting parents, on a constant dynamics of feedback and direction. Seibold et al. (2009) sharply alert that Millennials may not evidently realize that part of the effectiveness of self-managed teamwork lies in the control exerted by members within the group, so that the potential benefits of teamwork dynamics may transfigure into negative impacts. We then call out for this teamwork fallacy, as the above-described type of control emerges when team members collectively develop their own control system, establishing subgroups within the group itself, where control is negotiated and revealed per formal and informal team interaction, leading members to the growth of a collective sense of responsibility for the team's success. Hence, this dynamics may deconstruct the team spirit and dynamics as specific members may believe they are empowered and responsible to conquer compliance from other members, thus generating a subsystem of informal hierarchy, which may result in the overall team members' conformation to mutually agreed to norms (Barker 1993). Moreover, taking into consideration that literature commonly suggests that Millennials are excel followers (Howe and Strauss 2003), based on the argumentation above we can suggest that they can be more susceptible to this type of pressure. Nevertheless, we recognize room for discussion in regards to this idea as literature also states that Millennials are overly self-confident and individualistic (Pew Research Center 2007; Twenge 2009), thus possibly more resistant to these forms of control in the workgroup. Management will need to assess how these characteristics translate into workgroup conformity, dissecting as well its impact on team spirit and on the individual employee well-being and performance.

(d) **Reconstructing the Organizational Ladder**

Tulgan (2009) stresses that leaders need to develop into Millennial followers the sense of context, helping them understanding their position in the organization structure, aspect whose relevance is highlighted taking into consideration the fact that their short tenure can often be a source of frustration facing with older employees. At the same time, leaders shall acknowledge and respect that Millennials are making a contribution to the organization, often showing them clearly how even tasks in lower positions foster the organization's success and sustainability. We claim this to be a very important organizational practice in the current era, once Millennials, as above explained, are very demanding in terms of feedback and direction, driven by a very competitive mindset and moved towards the achievement of high standard goals, based on a pay-off mindset. Important to keep in mind to understand this argumentation is the fact that Millennials have been encouraged by their dedicated parents to challenge authority and to assert themselves, asking for preferential treatment when they believe they can get it, even in the work context (Howe and Strauss 2007).

Expectations of this type may be better understood if we consider they can be grounded on Millennials' socialization pattern, who has imprinted on them the idea of not being intimidated by individuals who are senior in age or status, once as children they were stimulated to establish connections with parents and friends of their parents (Howe and Strauss 2007), and as teens assumed comfortable expressing adults their ideas, demanding credibility despite their young age and lack of experiences (Tapscott 1998).

Consequently, their expectations for frequent, supportive and open communication, joined with their lack of formality in what concerns hierarchy respect, may cause senior-level employees/leaders to feel disrespected by these workers whom they believe not yet to have conquered such consideration. In fact, in these senior-level employees'/leaders' minds is the fact that Millennials may not yet fully acknowledge that amplified communication and access to increasing knowledge is associated with augmented responsibility.

On the other hand, and based on Tulgan's (2009) premise that Millennials do not define success by progressively climbing the corporate ladder, many organizations have chosen to avoid potential problems with tenure, currently raised mainly by Millennials, by implementing career progression based on meritocracy, determined by the quality of the work employees perform, not their time in a specific position or the academical degrees they may possess (Alsop 2008).

Millennials are impelled for more immediate rewards—not necessarily cash, as they are driven to work on projects that may offer them learning and growth opportunities right away (Baldonado and Spangenburg 2009). To better understand this argumentation, attention shall as well be shed on the fact that Millennials, who witnessed Boomer and Generation X parents endure the first wave of major layoffs, with it facing deep life challenges such as divorce or immigration, have interiorized lifetime employment as a myth (Tulgan 2009), hence not being concerned with what role they fulfil by moving through the organization's structure. Rather, they focus on the role the company will play in their life story. In order to effectively engage Millennials, Tulgan (2009) proposes that leaders need to explain and objectively demonstrate how working on a given project/position offers a chance for Millennials to acquire new knowledge and develop competencies, being it on the organization itself (thus evidencing how their work adds to the organizational value chain), or at least how working on such project/position provides Millennials a valuable set of capabilities to cite when applying for future opportunities.

(e) **Feedback and Information sharing: the Holy Grail**

Communication is the chief element in Millennials' workplace flows, prioritizing relationships as the core of their work experience once, as explained in previous sections, they tend to attach to people and not to ideologies (Brogan and Smith 2009). Hershatter and Epstein (2010) highlight that Millennials' managers consequently often describe such employees as "high-maintenance" or "needy", so that managers may frequently see themselves having to spend an uneven amount of time managing people who were supposed to have been hired to help them. Nevertheless, positive outcomes can emerge from this dynamic, boosting organizational performance, but to achieve such point managers shall not hinder Millennial's needs, they shall accompany at best possible these workers, inciting them to progressively take ownership of tasks and responsibilities, coming into a point in which the relationship is so intensely developed that they feel not only technically ready to act on their own but also comfortable to make their own journey and to be fully accountable for it.

While these may seem overwhelming demands to charge on a workplace, to better contextualize it we need to recall that throughout their lives Millennials have been encouraged to maintain similarly close relationships with parents, teachers and mentors, thus, they are likely to expect an analogous treatment from their supervisors. Indeed, Millennials pursue ample feedback because it assures they are continually moving along a progressive path towards the achievement of their desired outcomes, as they have been indoctrinated from their earliest moment to seek approval and affirmation. Barnes (2009) provides a useful input noting that silence is often perceived as a negative response by this generation: as these employees are motivated by the desire to constantly please and seek praise from their supervisors, they want to know when they have done so and how to improve (Smith and Galbraith 2012). Apropos, Alessandra's (1995: 23) input is very interesting, as the author set forth the "Platinum

Rule": whereas the "Golden Rule" focuses on treating others the way you want to be treated, the "Platinum Rule" suggests to "do unto others the way they want to be done unto", crystalizing a communication approach that focusses more on the receiver's perspective "by learning how to do what they want done", establishing an "instant rapport".

Subsequently, Tulgan (2009) contends that leaders who want to recruit and retain the most talented Millennials will have to adjust how and how often they deliver feedback. On the same logics, Burkus (2010) and Alsop (2008) claim that frequent and informal feedback shall occur—quarterly, monthly or after a project's completion, because it will incite involvement and, afterwards, organizational attachment, because, following Myers and Oetzel (2003), Millennials perceive they are becoming recognized and appreciated. At the same time, this may encourage such workers to realize the value of investing time developing trust among coworkers, an important aspect if we consider their speedy rhythm of life and short time expectation frame to achieve the desired outcomes (Myers and Sadaghiani 2010).

(f) **Reconceiving Performance Appraisal**

Millennials seem more propense to operate within existing structures, expecting to rely on these to provide them with the resources needed to achieve the desired outcomes. Therefore, they also expect that the strong structural framework and assessment systems in which they were raised continue in their work life, this grounded on a combination of youth's idealism and the sheltering protection they have been afforded by their parents (Brack and Kelly 2012; Hershatter and Epstein 2010). Thus, the performance appraisal process needs to reflect the employees' contributions rather than merely provide a list of their efforts, thus, shall emphasize outcomes in terms of specific objectives, avoiding *The Micromanagement Fallacy.* Therefore, organizations shall encourage Millennials' desire to contribute and reward them with opportunities for advancement, building a sense of loyalty, retaining them in the workforce and taking active steps to develop them as future leaders (Beekman 2011).

Classical project management has been based on control mechanisms that strive for accuracy, productivity, determinism, optimization and top-down execution; however, nowadays, it is increasingly proposed a step-by-step methodology favouring rapid changes or midcourse corrections (Thomas and Mullaly 2007), methodologies clearly more appealing to Millennials once the world in which they inhabit increasingly reflects a similar environment due to the digital realm's triumph with Globalization. This reveals how the application of simple principles wires the cooperative, participative and dispersed leadership, because at its basis grounds the concepts of agility, speed, simplicity and readiness for further processing/incorporation, thus providing active inputs towards the tying up innovation, creativity, knowledge creation and sharing (Rising and Janoff 2000).

Being Millennials a generation of highly competitive people used to be rewarded and receive constant feedback, strongly sheltered and nurtured by their doting parents, it is crucial that their performance appraisal process concentrates on highlighting positive contributions rather than pointing failures. Nonetheless, failures shall be evidenced to Millennials, and in doing so we suggest the relevance of framing them as a positive learning experience and providing objective examples so that Millennials can clearly and quickly understand the issue, also fostering creative alternatives to meet organizational goals. So, the performance appraisal process should be designed in order to provide specific and objective reporting and suggestions: this way, Millennials would be driven towards a chance to improve their own skills, feeding their competitive and achieving mindset, positively affecting the organizational performance as such approach would inspire them to outspread their efforts beyond the basic dynamics of simple task completion, which would, in turn, foster these employees' retention potential. Thus, it would locate and evaluate the employee's actions within a wider context: it would indeed not only allow a focus on individual efforts but also consider how the employee contributed as a team member—we shall recall Millennials' natural propension to work in groups.

7 Conclusion

As previous generations retire, Millennials are growingly assuming themselves as the future of organizational dynamics, not only because they currently represent the most recent and numerous generation in the workforce, but as well due to their specific characteristics and interaction style. Based on this premise, and via a thorough literature review, this study assesses the construct of Millennials and their leadership and followership attributes through an analysis based on inputs from several scientific domains, discussing how these contribute to the redefined dynamics this generation impels at the organizational level. Highlighting and contextualizing Millennials traits and how these reflect into the organizational structures, practices and processes, this work sheds light on the challenges organizations may face in the current era due to the increasing inclusion of this cohort into the workforce, as multigenerational challenges arise once for the first time four different generations of workers coexist within the same organization—namely Veterans, Baby Boomers, Generation X and Millennials. The idiosyncrasies among these groups may have an impact on how leaders interact with them, as differences can create problems among team members that eventually reduce individual and organizational performance. Thus, this work provides inputs aiming at retaining Millennials long enough in the organization to develop them into strategic and valuable assets, fostering their followership in a way to mature and to incite them to be its future leaders.

Millennials' psychodynamics require management attention and planning, foremost because many of these are based on the carry-over of some traits deeply imprinted in their parents' mindset, consequently, we claim that the "helicopter parenthood" (Twenge 2009) developed into Millennials some aspects this generation then transposes into the workplace dynamics, crystallizing into a "Trophy Kids" mentality (Alsop 2008). As so, their early life experiences climaxed into a generation that, we claim, is driven by a "pay-off" mindset, always motivated by achieving, progressing and succeeding, deeply constructed at the basis of a competitive moto, at the same time constantly needing guidance and feedback in order to averse risk and to best track their learning path, thus feeling sheltered, nurtured, appreciated and validated not only by peers but (more importantly) by their superiors, being them parents, mentors, managers, teachers, etc.

We suggest, then, that organizations will need to develop communication plans to guide Millennials in the direction the organization requires to foster workforce cohesion. Leaders should as well direct efforts towards the relationships they form with Millennials in the workplace, based on the argument that this generation seeks more meaning and value from these relationships than other generational cohorts in the workplace. Based on the premise that their educational path and other early life experiences may not prepare Millennials for successful entry into a workplace shaped by previous generations, this work advises that leadership styles that support participation and collaboration via shared responsibilities seem to be more attractive to this generation in terms of work dynamics.

Based on the argument that Millennials operate through a knowledge-intensive, highly relational and networked dynamics, this work advocates the positive impact of implementing coaching approaches to best manage the current multigenerational workforce that marks todays' organizational structures. We conclude that future organizational paradigms need to develop a multigenerational cooperative and concerted culture: by providing Millennials with meaningful and enhancing experiences, objectively showing them their contribution and respecting it, facing them with challenges on a continuous learning and skills-improvement progression, implementing mentoring, providing ongoing feedback and adopting flexible dynamics, reconsidering the performance appraisal process and fluidizing organizational structures, thus plasticizing organizational mobility. This way, leaders would be able to retain Millennials as followers, who will (also and at a later stage) develop into future organizational leaders. The first step would be acknowledging that organizational structures, modes of personal engagement and work processes shall adapt on an enduring basis, not only to best accommodate this cohort into the workforce, capitalizing on its talent but as well to best manage the challenges posed by a multigenerational workforce. Expanding such logics, this work then recommends the relevance of implementing redesigned organizations concentrating on people, communication, contacts, innovation and creativity, requiring refined collective dynamics, learning and experimenting processes through relationally driven and technologically cooperative and concerted processes, fluidizing the hierarchical structure and implementing new manifestations of leadership that recreate, flex, learn, adapt and serve.

As Millennials will continue to join the workforce, we can clearly state the relevance of this work: by understanding Millennials' leadership style, it provides further inputs to design a work environment where leadership effectiveness can be maximized, which in turn may have positive outcomes in fostering individual, group and organizational performance. By understanding the followership style of Millennials, it may as well provide insights so that organizations are able to best manage Millennials work performance capitalizing on their talent at more efficient levels. The symbiosis of both inputs provides a more accurate and efficient management of the current multigenerational workforce challenges. This work has, nonetheless, limitations, which we present as an impulse to foster research on the topic. Indeed, it mainly concentrates on the beliefs, aspirations, attitudes and values of Millennials in the workplace identified as stated by previous research and how it can affect this generation's leadership and followership styles in the workplace, thus, it is discussible that there might be variables not considered in the analysis, as, for example, it has been shown that individual (Valliant and Loring 1998) along with organizational (Lok and Crawford 2004; Ogbonna and Harris 2000) factors may affect leadership and followership styles. Future research should as well seek more thoroughly to understand Millennials' sources of leadership socialization and the values this generation communicates once achieving leadership roles. Also, it shall empirically examine the impact of the multigenerational workforce challenges on individual and organizational performance.

References

Alessandra, T. (1995). The platinum rule. *Vital Speeches of the Day, 62*(1), 23–27.
Alsop, R. (2008). *The trophy kids group up: How the Millennial generation is shaping up the workplace*. San Francisco: Jossey-Bass.
Alsop, R., Nicholson, P., & Miller, J. (2009). Gen Y in the workforce commentary. *Harvard Business Review, 87*(2), 43–49.
Altizer, T. E. (2010). Motivating Gen Y amidst global economic uncertainty. *Journal of Learning in Higher Education, 6*(1), 44–54.
Avolio, B. J., Walumbwa, F. O., & Weber, T. J. (2009). Leadership: Current theories, research, and future directions. *Annual Review of Psychology, 60,* 421–449.
Balda, J. B., & Mora, F. (2011). Adapting leadership theory and practice for the networked, millennial generation. *Journal of Leadership Studies, 5*(3), 13–24.
Baldonado, A., & Spangenburg, J. (2009). Leadership and the future: Gen Y workers and two-factor theory. *Journal of American Academy of Business, 15*(1), 99–103.
Barker, J. R. (1993). Tightening the iron cage: Concertive control in self-managing teams. *Administrative Science Quarterly, 38,* 408–437.
Barnes, G. (2009). Guess who's coming to work: Generation Y Are you ready for them? *Public Library Quarterly, 28*(1), 58–63.
Barzilai-Nahon, K., & Mason, R. (2010). How executives perceive the net generation. *Information, Communication and Society, 13*(3), 396–418.
Beekman, T. (2011). Fill in the generation gap. *Strategic Finance, 93*(3), 15–17.
Bjugstad, K., Thach, E. C., Thompson, K. J., & Morris, A. (2006). A fresh look at followership: A model for matching followership and leadership styles. *Journal of Behavioral and Applied Management, 7*(3), 304–319.
Brack, J., & Kelly, K. (2012). Maximizing millennials in the workplace. *UNC Executive Development, 22*(1), 2–14.
Broadbridge, A., Maxwell, G., & Ogden, S. (2007). Experiences, perceptions and expectations of retail employment for Generation Y. *Career Development International, 12,* 523–544.
Brogan, C., & Smith, J. (2009). *Trust agents: Using the web to build influence, improve regulation, and earn trust*. Hoboken: Wiley.
Brooks, D. (2001). The organization kid. *The Atlantic Monthly, 287*(4), 40–54.
Bryman, A. (1992). *Charisma and leadership in organisations*. Newbury Park: Sage.

Burkus, D. (2010). Developing the next generation of leaders: How to engage s in the workplace. *Leadership Advance Online, 14,* 1–6.

Burns, J. M. (1978). *Leadership.* New York: Harper & Row Publishers.

Campione, W. A. (2015). Corporate offerings: Why aren't millennials staying? *Journal of Applied Business and Economics, 17*(4), 60–75.

Caraher, L. (2015). *Millennials and management: The essential guide to making it work at work.* Brookline: Bibliomotion.

Carless, S. A., & Wintle, J. (2007). Applicant attraction: The role of recruiter function, work-life balance policies and career salience. *International Journal of Selection and Assessment, 15*(4), 394–404.

Cennamo, L., & Gardner, D. (2008). Generational differences in work values, outcomes, and person-organisation values fit. *Journal of Managerial Psychology, 23*(8), 891–906.

Chen, Y. F., & Tjosvold, D. (2006). Participative leadership by American and Chinese managers in China: The role of relationships. *Journal of Management Studies, 43*(8), 1727–1752.

Chou, S. Y. (2012). Millennials in the workplace: A conceptual analysis of s' leadership and followership styles. *International Journal of Human Resource Studies, 2*(2), 71–83.

Cisco Corporation. (2011). *Cisco connected world technology report.* Retrieved on the 27th February 2016, from http://www.cisco.com/c/en/us/solutions/enterprise/connectedworld-technology-report/index.html#~2011.

Cole, G., Smith, R., & Lucas, L. (2002). The debut of Generation Y in the American workforce. *Journal of Business Administration Online, 1*(2), 1–10.

Coomber, B., & Barriball, K. (2007). Impact of job satisfaction components on intent to leave and turnover for hospital-based nurses: A review of the research literature. *International Journal of Nursing Studies, 44*(2), 297–314.

Costanza, D. P., Badger, J. M., Fraser, R. L., Severt, J. B., & Gade, P. A. (2012). Generational differences in work-related attitudes: A meta-analysis. *Journal of Business and Psychology, 27*(4), 375–394.

Credo, K. R., Lanier, P. A., Matherne, C. F., III, & Cox, S. S. (2016). Narcissism and entitlement in millennials: The mediating influence of community service self-efficacy on engagement. *Personality and Individual Differences, 101,* 192–195.

Crumpacker, M., & Crumpacker, J. M. (2007). Succession Planning and Generational Stereotypes: Should HR Consider Age-Based Values and Attitudes a Relevant Factor or a Passing Fad? *Public Personnel Management, 36*(4), 349–369.

Dansereau, F., Graen, G., & Haga, W. J. (1975). A vertical dyad linkage approach to leadership within formal organizations: A longitudinal investigation of the role-making process. *Organizational Behavior and Human Performance, 13*(1), 46–78.

DePree, M. (1989). *Leadership is an art.* New York: Doubleday Business.

Dulin, L. (2008). Leadership preferences of a generation y cohort. *Journal of Leadership Studies, 2*(1), 43–59.

Dwyer, R. (2009). Prepare for the impact of the multi-generational workforce! *Transforming Government: People, Process and Policy, 3*(2), 101–110.

Egri, C. P., & Ralston, D. A. (2004). Generation cohorts and personal values: A comparison of China and the U.S. *Organization Science, 15*(2), 210–220.

Emeagwali, N. S. (2011). Millennials: Leading the charge for change. *Techniques: Connecting Education and Careers, 86*(5), 22–26.

Erickson, T. (2008). *Plugged in: The Generation Y guide to thriving at work.* Boston: Harvard Business Press.

Espinoza, C., Ukleja, M., & Rusch, C. (2010). *Managing the millennials: Discover the core competencies for managing today's workforce.* Hoboken: Wiley.

Fiedler, F. E. (1967). *A theory of leadership effectiveness.* New York: McGraw-Hill.

Gibson, W. J., Greenwood, R. A., & Murphy, E. F. (2010). Analyzing generational values among managers and non-managers for sustainable organizational effectiveness. *SAM Advanced Management Journal, Winter*, 33–43.

Gilley, A., Waddell, K., Hall, A., Jackson, S. A., & Gilley, J. W. (2015). Manager behavior, generation, and influence on work-life balance: An empirical investigation. *Journal of Applied Management and Entrepreneurship, 20*(1), 3–23.
Goldthorpe, J., Lockwood, D., Bechhofer, F., & Platt, J. (1968). *The affluent worker: Industrial attitudes and behaviour*. Cambridge: Cambridge University Press.
Gorman, P., Nelson, T., & Glassman, A. (2004). The Millennial generation: A strategic opportunity. *Organizational Analysis, 12*(3), 255–270.
Graen, G. B., & Scandura, T. A. (1987). Toward a psychology of dyadic organizing. *Research in Organizational Behavior, 9,* 175–208.
Graen, G. B., & Uhl-Bien, M. (1995). Relationship-based approach to leadership: Development of leader-member exchange (LMX) theory of leadership over 25 years: Applying a multi-level multi-domain perspective. *Leadership Quarterly, 6,* 219–247.
Greenberg, E. H., & Weber, K. (2008). *Generation We: How Millennial youth are taking over America and changing our world forever*. Emeryville: Pachatusan.
Greenfield, P. M. (1998). The cultural evolution of IQ. In U. Neisser (Ed.), *The rising curve: Long-term gains in IQ and other measures* (pp. 81–123). Washington, D. C.: American Psychological Association.
Gursoy, D., Maier, T. A., & Chi, C. G. (2008). Generational differences: An examination of work values and generational gaps in the hospitality workforce. *International Journal of Hospitality Management, 27*(3), 448–458.
Gursoy, D., Chi, C. G., & Karadag, E. (2013). Generational differences in work values and attitudes among frontline and service contact employees. *International Journal of Hospitality Management, 32,* 40–48.
Harris-Boundy, J., & Flatt, S. J. (2010). Cooperative performance of Millennials in teams. *Review of Business Research, 10*(4), 30–46.
Hartman, J. H., & McCambridge, J. (2011). Optimizing Millennials' Communication Styles. *Business Communication Quarterly, 74*(1), 22–44.
Helyer, R., & Lee, D. (2012). The 21st century multiple generation workforce: Overlaps and differences but also challenges and benefits. *Education + Training, 54*(7), 545–578.
Hershatter, A., & Epstein, M. (2010). Millennials and the world of work: An organization and management perspective. *Journal of Business and Psychology, 25*(2), 211–223.
Hewlett, S., Sherbin, L., & Sumberg, K. (2009). How Gen Y and Boomers will reshape your agenda. *Harvard Business Review, 87*(7/8), 71–76.
Hodson, R. (2004). Demography or respect? Work group demography versus organizational dynamics as determinants of meaning and satisfaction at work. *British Journal of Sociology, 53*(2), 291–317.
Hole, D., Zhong, L., & Schwartz, J. (2010). Talking about whose generation. *Deloitte Review, 6*(1), 83–97.
Howe, N., & Strauss, W. (2000). *Millennials rising: The next great generation.* New York: Vintage Books.
Howe, N., & Strauss, W. (2003). *Millennials go to college: Strategies for a new generation on campus: Recruiting and admissions, campus life, and the classroom.* Washington, D. C.: AACRAO.
Howe, N., & Strauss, W. (2007). *Millennials and K-12 Schools: Educational Strategies for a New Generation.* Great Falls: LifeCourse Associates.
Howell, J. P., & Costley, D. L. (2001). *Understanding behaviors for effective leadership.* Upper Saddle River: Prentice Hall.
Hulin, C. L., & Judge, T. A. (2003). Job attitudes: A theoretical and empirical review. In W. C. Borman, D. R. Ilgen, & R. J. Klimoski (Eds.), *Handbook of psychology*, vol. 12, (pp. 255–276). Wiley: Hoboken.
Huntley, R. (2006). *The World According to Y: Inside the New Adult Generation.* Crows Nest: Allen and Unwin.

Inglehart, R. F. (2008). Changing values among Western publics from 1970 to 2006. *West European Politics, 31*(1–2), 130–146.

Ismail, K. M., & Ford, D. L. (2010). Organizational leadership in Central Asia and the Caucasus: Research considerations and directions. *Asia Pacific Journal of Management, 27*(2), 321–340.

Jacobson, W. S. (2007). Two's company, three's a crow, and four's a lot to manage: Supervising in today's intergenerational workplace. *Popular Government, 17*(Fall), 18–23.

Kaifi, B., Nafei, W., Khanfar, N., & Kaifi, M. (2012). A multi-generational workforce: Managing and understanding s. *International Journal of Business and Management, 7*(24), 88–93.

Lancaster, L. C., & Stillman, D. (2010). *The M-factor: How the Millennial Generation is rocking the workplace.* New York: HarperCollins.

Lok, P., & Crawford, J. (2004). The effect of organisational culture and leadership style on job satisfaction and organisational commitment. *Journal of Management Development, 23*(4), 321–338.

Lythcott-Haims, J. (2015). *How to raise an adult: Break free of the overparenting trap and prepare your kid for success.* New York: Henry Holt and Company.

Mannheim, K. (1952). The problem of generations. In K. Mannheim (Ed.), *Essays on the sociology of knowledge* (pp. 276–320). London: Routledge and Kegan Paul.

Marston, C. (2007). *Motivating the "What's in it for me?" workforce: Manage across the generational divide and increase profits.* Hoboken: Wiley.

Marston, C. (2009). *Myths about Millennials: Understand the myths to retain Millennials.* Retrieved on the 18th July 2020, from http://humanresources.about.com/od/managementtips/a/millennial_myth.htm.

Martin, C. A. (2005). From high maintenance to high productivity: What managers need to know about Generation Y. *Industrial and Commercial Training, 37*(1), 39–44.

McGonagill, G., & Pruyn, P. W. (2010). *Leadership development in the U.S.: Principles and patterns of best practice.* Berlin: Bertelsmann Stiftung Leadership Series, S. Vopel.

McGuire, D., By, R. T., & Hutchings, K. (2007). Towards a model of human resource solutions for achieving intergenerational interaction in organizations. *Journal of European Industrial Training, 31,* 592–608.

Meindl, J. R. (1995). The romance of leadership as a follower-centric theory: A social constructionist approach. *The Leadership Quarterly, 6*(3), 329–341.

Murphy, W. M. (2012). Reverse mentoring at work: Fostering cross-generational learning and developing millennial leaders. *Human Resource Management, 51*(4), 549–573.

Mushonga, S. M., & Torrance, C. G. (2008). Assessing the relationship between followership and the big five factor model of personality. *Review of Business Research, 8*(6), 185–193.

Myers, K. K., & Oetzel, J. G. (2003). Exploring the dimensions of organizational assimilation: Creating and validating a measure. *Communication Quarterly, 51*(4), 438–457.

Myers, K. K., & Sadaghiani, K. (2010). Millennials in the workplace: A communication perspective on Millennials' organizational relationships and performance. *Journal of Business Psychology, 25*(2), 225–238.

Ng, E. S., Schweitzer, L., & Lyons, S. T. (2010). New Generation, Great Expectations: A Field Study Of The Millennial Generation. *Journal of Business and Psychology, 25*(2), 281–292.

Niemiec, S. (2000). Finding Common Ground for All Ages. *SDM: Security Distributing and Marketing, 30*(3), 81.

Nolan, L. S. (2015). The roar of millennials: Retaining top talent in the workplace. *Journal of Leadership, Accountability and Ethics, 12*(5), 69–75.

Ogbonna, E., & Harris, L. C. (2000). Leadership style, organizational culture and performance: empirical evidence from UK companies. *International Journal of Human Resource Management, 11*(4), 766–788.

Oppel, W. A. (2007). Generational diversity: the future of the American workforce. *Leadership Advance Online, 9,* 1–3.

Özçelik, G. (2015). Engagement and retention of the millennial generation in the workplace through internal branding. *International Journal of Business and Management, 10*(3), 99–107.

Palfrey, J., & Gasser, U. (2008). *Born digital*. New York: Basic Books.

Pew Research Center. (2007). *How young people view their lives, futures, and politics: A portrait of "Generation Next"*. Retrieved on the 18th July 2020, from http://people-press.org/report/300/a-portrait-of-generation-next.

Pfeffer, J., & Salancik, G. R. (1975). Determinants of supervisory behavior: A role set analysis. *Human Relations, 28*(2), 139–154.

Pînzaru, F., Vătămănescu, E. M., Mitan, A., Săvulescu, R., Vițelar, A., Noaghea, C., et al. (2016). Millennials at work: Investigating the specificity of generation Y versus other generations. *Management Dynamics in the Knowledge Economy, 4*(2), 173–192.

Prensky, M. (2001). Digital natives, digital immigrants. *On the Horizon, 9*(5), 1–6.

Pyöriä, P., Ojala, S., Saari, T., & Järvinen, K.-M. (2017). The millennial generation: a new breed of labour? *SAGE Open, 7*(1), 1–14.

Raines, C. (2002). *Connecting generations: The sourcebook for a new workplace*. Berkeley: Crisp Publications.

Rikleen, L. S. (2011). *Creating tomorrow's leaders: The expanding roles of millennials in the workplace*. Boston: Boston College Center for Work and Family.

Rising, L., & Janoff, N. S. (2000). The Scrum software development process for small teams. *IEEE Software, 17*(4), 26–32.

Rosen, L. D. (2007). *Me, MySpace and I: Parenting the net generation*. Hampshire: Palgrave Macmillan.

Ryder, N. B. (1965). The cohort as a concept in the study of social change. *American Sociological Review, 30*(6), 843–861.

Sadaghiani, K., & Myers, K. K. (2009). *Parents' influence on leadership values: The vocational anticipatory socialization of young millennial adults*. Paper presented at the Western States Communication Association, 80th Annual Convention, Mesa, Arizona.

Salahuddin, M. M. (2010). Generational differences impact on leadership style and organizational Success. *Journal of Diversity Management, 5*(2), 1–6.

Scroggins, W. A. (2008). The relationship between employee fit perceptions, job performance, and retention: Implications of perceived fit. *Employee Responsibilities and Rights Journal, 20*(1), 57–71.

Seibold, D. R., Kang, P., Gailliard, B. M., & Jahn, J. L. S. (2009). Communication that damages teamwork: The dark side of teams. In P. Lutgen-Sandvik & B. Davenport Sypher (Eds.), *Destructive organizational communication: Processes, consequences, and constructive ways of organizing* (pp. 267–289). New York: Routledge.

Small, G., & Vorgan, G. (2008). *iBrain: Surviving the technological alteration of the modern mind*. New York: Harper Collins.

Smith, K. T. (2010). Work-life balance perspectives of marketing professionals in generation Y. *Services Marketing Quarterly, 31*(4), 434.

Smith, S. D., & Galbraith, Q. (2012). Motivating millennials: Improving practices in recruiting, retaining, and motivating younger library staff. *The Journal of Academic Librarianship, 38*(3), 135–144.

Smith, T. J., & Nichols, T. (2015). Understanding the millennial generation. *The Journal of Business Diversity, 15*(1), 39–47.

Spiegal, D. (2011). *Why hiring Millennials is good for your business*. Open Forum. Retrieved on the 13th June 2020, from http://www.openforum.com/articles/why-hiring-millennials-is-good-for-your-business.

Stewart, J. S., Oliver, E. G., Cravens, K. S., & Oishi, S. (2017). Managing millennials: Embracing generational differences. *Business Horizons, 60*(1), 45–54.

Tapscott, D. (1998). *Growing up digital: The rise of the net generation*. New York: McGraw-Hill.

Tapscott, D. (2009). *Grown up digital: How the net generation is changing your world*. New York: McGraw-Hill.

Thomas, J., & Mullaly, M. (2007). Understanding the value of project management: First steps on an international investigation in search of value. *Project Management Journal, 38*(3), 74–89.

Thompson, C., & Gregory, J. B. (2012). Managing millennials: A framework for improving attraction, motivation, and retention. *The Psychologist-Manager Journal, 15*(4), 237–246.

Timo, N., & Davidson, M. (2005). A survey of employee relations practices and demographics of Mnc chain and domestic luxury hotels In Australia. *Employee Relations, 27*(2), 175–192.

Tolbzie, A. (2008). Generational differences in the workplace. *Research and Training Center of Community Living, 19,* 1–13.

Tourangeau, A. E., & Cranley, L. A. (2006). Nurse intention to remain employed: Understanding and strengthening determinants. *Journal of Advanced Nursing, 55*(4), 497–509.

Trzesniewski, K. H., & Donnellan, M. B. (2010). Rethinking "Generation Me": A study of cohort effects from 1976-2006. *Perspectives on Psychological Science, 5*(1), 58–75.

Tulgan, B. (2009). *Not everyone gets a trophy: how to manage generation Y*. San Francisco: Jossey-Bass.

Twenge, J. M. (2009). Generational changes and their impact in the classroom: Teaching generation me. *Medical Education, 43*(5), 398–405.

Twenge, J. M. (2013). The evidence for Generation Me and against Generation We. *Emerging Adulthood, 1*(1), 11–16.

Twenge, J., & Campbell, W. K. (2001). Age and birth cohort differences in self-esteem: A cross temporal meta-analysis. *Personality and Social Psychology Review, 5*(4), 321–344.

Twenge, J. M., & Campbell, S. M. (2009). *The narcissism epidemic: Living in the age of entitlement*. New York: Free Press.

Twenge, J. M., Campbell, S. M., Hoffman, B. J., & Lance, C. E. (2010). Generational differences in work values: Leisure and extrinsic values increasing, social and intrinsic values decreasing. *Journal of Management, 36*(5), 1117–1142.

Underwood, C. (2007). *The generational imperative—Understanding generational differences in the workplace, marketplace and living room.* South Carolina: BookSurge.

Valliant, P. M., & Loring, J. E. (1998). Leadership style and personality of mock jurors and the effect on sentencing decisions. *Social Behavior and Personality: an International Journal, 26*(4), 421–424.

Zemke, R., Raines, C., & Filipczak, B. (2000). *Generations at work: Managing the clash of veterans, boomers, xers and nexters in your workplace.* New York: AMACOM.

The Dark Side of Human Resources Management: The Perceptions of Different Organizational Actors

João Leite Ribeiro and Delfina Gomes

> Sometimes I wonder whether the world is being run by smart people who are putting us on or by imbeciles who really mean it.
> —Mark Twain

Abstract Fundamental as it is that HR managers work as team leaders and potenciate the creation of friendly and stimulating environments, increasing attention has been given lately to malpractices of Human Resources Management (HRM) by researchers, organizations and public opinion. As part of a broader project that aims to understand the perceptions and the attributions of different types and situations of malpractices in HRM, this aims to understand how the way in which the position of HR Director is exercised, and its influences on the performance of other management positions, can give rise to poor HRM strategies, behaviors, and situations. In this study, 60 interviews were conducted with organizational actors, and Grounded Theory was applied to analyze the data. The interviewees were employed at 3 Portuguese companies and held different hierarchical positions: peers of HR managers and employees from different organizational functions. This study contributes to unveiling the dark side of HRM. The findings highlight that the little respect for the values and the demagogy of the discourse about the importance of people as the most valuable capital of an organization leads, in face of what is practiced on a daily basis, to a diminishing of HRM, the HR department, managers and the organization. The study calls for a deep understanding of those malpractices in order to allow the flourishing of best practices in HRM, in particular for stimulating creative and innovative teams, such as the creation of a team coaching staff in organizations.

J. L. Ribeiro (✉)
School of Economics and Management, Interdisciplinary Centre of Social Sciences (CICS.NOVA. UMinho), University of Minho, Braga, Portugal
e-mail: joser@eeg.uminho.pt

D. Gomes
School of Economics and Management, Research Center in Political Science, University of Minho, Braga, Portugal

C. Machado and J. P. Davim (eds.), *Coaching for Managers and Engineers*, Management and Industrial Engineering,
https://doi.org/10.1007/978-3-030-71105-4_6

1 Introduction

In an ideal world leaders, including Human Resources (HR) Directors, would stimulate their teams and company collaborators by creating emotionally safe conditions and opportunities for those to express their intrinsic motivation (Graen et al. 2020). The authors Graen et al. (2020) propose a set of best practices for HRM, in particular for stimulating creative and innovative teams. Among these practices is the creation of a team coaching staff responsible to deliver advanced practices in projects (Graen et al. 2020). However, to achieve this goal for best practices in Human Resources Management (HRM), it is still necessary to unveil and debate malpractices that persist in the daily life of companies.

Although issues and behaviors related with malpractices of HRM are not a novelty within HRM research, the fact is that increasing attention has been given lately to this topic by researchers, organizations, and public opinion (Crawshaw 2009; Einarsen et al. 2003; Eurofound 2014; Fahie and Devine 2014; Martinko et al. 2013; Tepper 2007).

The issues raised by malpractices in HRM confirms and reinforces the distance between rhetoric and practice in the value attributed to persons and their management. This gap happens in all kinds of organizations and manifests itself through a wide range of practices and organizational internal realities (Bernotaute and Malinauskiene 2017; Eurofound 2014; Hershcovis 2011; Legge 1995; Morten and Einarsen 2018).

The present study is part of a broader project that aims to understand the perceptions and the attributions of different types and situations of malpractices in the domain of management in general and in HRM in specific. The project addresses the following research question: What type, nature, attributions, and enhancers of malpractices in HRM different organizational actors of diverse hierarchical and functional levels perceive in their organizations?

In this study, the objective is to understand how the way in which the position of HR Director is exercised, and its influences on the performance of other management positions, can give rise to poor HRM strategies, behaviors, and situations. Thus, this study contributes to unveiling the dark side of HRM (Einarsen et al. 2019; Guest 2017; Krasikova et al. 2013; Tepper et al. 2017).

To develop this study, 60 interviews with different organizational actors were used, and Grounded Theory was applied to analyze the data. The interviewees were employed at three Portuguese companies and represented different hierarchical positions: peers of HR managers and employees from different organizational functions.

It was intended to address through the interviews the opinions, perceptions regarding strategies, behaviors, and situations that, from the perspective of different actors from different companies, were perceived as generators/enhancers of less appropriate or very inadequate behaviors and situations in relation to HRM. The analysis of the data highlight aspects such as lack of consistency between values, principles, policies, and daily practices. This lack of coherence was explicitly contextualized in the lack of concrete action by HRM specialists—director, technicians, and other employees of the HR Department—but also, and mainly, in some peer directors of the HR director and middle and line managers who take on professional behaviors perceived as examples of what should never happen in HRM.

The remainder of this chapter is structured as follows. The next section presents a brief outline of the literature on malpractices of HRM. The methodology of the study and a description of the participants are presented in the subsequent section. Next, different organizational actors' opinions and perceptions regarding strategies, behaviors, and situations that were perceived as generators/enhancers of less appropriate or very inadequate behaviors and situations in relation to HRM are analyzed. The final section offers a discussion and conclusion.

2 Theoretical Underpinnings

Referring "malpractices in HRM" implies to understand that this subject is positioned in the realm of people's judgment and perceptions, particularly employees regardless of their hierarchical and functional level. Within this topic, and as argued by Tepper (2000, p. 178), a growing body of research has explored abusive supervision, "subordinates' perceptions of the extent to which supervisors engage in the sustained display of hostile verbal and nonverbal behaviors, excluding physical contact".

When studying malpractices in HRM the aim is to analyze problematical behaviors, such as mobbing; bullying; sexual and physical harassment; discrimination; depreciation of somebody's work; psychological terrorism; and victimization (Crawshaw 2009; Einarsen et al. 2003; Hershcovis 2011; Khoo 2010; Morten and Einarsen 2018). But it is also important to consider in these malpractices subtle behaviors, which are equally perverse and harmful to human dignity, such as criticism; intimidation; veiled threats; humiliation; exclusion; disruption of social interactions; disciplinary proceedings based on false accusations; public ridicule; lies about the person, configured in what is abusive supervision (Eurofound 2014; Khoo 2010; Tepper et al. 2006).

This is also a research topic where the possibility of personal, group and institutional exploitation must be considered, what calls for further research from different locations and cultures (Bernotaute and Malinauskiene 2017; Guest 2017; Harris et al. 2007; Kowalski and Loretto 2017; Morten and Einarsen 2018; Tepper 2007).

There is a dark side in management and in HRM that cause pain, various disorders, sleep disturbance, exhaustion, anxiety, depression, loss of self-esteem, stress, burnout, and even violent deaths (Aborde de Chatillon and Richard 2015; Astrauskaite et al. 2010; Bernotaute and Malinauskiene 2017; Daniels et al. 2017; Fahie and Devine, 2014; Hershcovis 2011; Tepper et al. 2017). This dark side is present in the news, such as the suicides, while others do not become public, laying under the iceberg tip. Increasing research on these issues is stimulated by the costs, financial and non-financial, for organizations caused by abusive behaviors. In a society ruled by numbers and statistics, it is easy to forget that living things bleed and in this specific case it is people we are talking about (Bernotaute and Malinauskiene

2017; Eurofound 2014; Harris et al. 2007; Morten and Einarsen 2018; Tepper 2007). As argued by Burton et al. (2014, p.871), "Understanding how employees come to believe they have been abused is important not only in light of the substantial social and financial costs of abuse but also because employees can vary considerably in how they perceive supervisors' actions".

Malpractices in HRM is a current, pertinent, sensitive, politically and socially incorrect theme, often hidden by shame, lack of practical proof, emotional or physical blackmail, and fear (Bernotaute and Malinauskiene 2017; Morten and Einarsen 2018). A theme concealed in conferences, premiered in lectures where one tries to give an almost perfect, if not idyllic, picture of organizational environments, where in reality there is a dark side of management that wants to be bleached, with more or less innovative forms of organizational eclipses or total darkness (Bernotaute and Malinauskiene 2017; Morten and Einarsen 2018). However, these malpractices have impact on employees and their perception of justice, with previous research showing "that perceptions of abusive supervision diminish employees' perceptions of justice" (Burton et al. 2014, p. 895; see also Bernotaute and Malinauskiene 2017; Burton and Hoobler 2011; Einarsen et al. 2019; Guest 2017; Khoo 2010; Teixeira et al. 2011; Tepper 2000).

The study of organizational cynicism and manipulative and abusive relationships has been focusing on organizational policies in general, and in particular HRM and leadership, namely their lack of adequacy and quality (Davis and Gardner 2004; Einarsen et al. 2019; Guest 2017).

Different aspects discredit holders of management positions mainly in the management of people and teams, such as lack of quality and clarity in the manager-worker relationship, along with role ambiguities and hierarchical and functional responsibilities; distance between managers and workers; deficient and not very transparent communication systems; organizational cynicism and the perceived lack of technical and practical knowledge and behavioral skills (Atwater et al. 2000; Davis and Gardner 2004; Einarsen et al. 2019; Guest 2017; Katz and Kahn 1978; Krasikova et al. 2013; Valle and Perrewe 2000).

3 Methodology

This study is part of a broader investigation developed under an interpretative paradigm by assuming that reality is a social construction and cannot be understood independently from the actors that create that reality (Urquhart 2013). The empirical study is based on qualitative research designed to understand phenomena through the meanings that individuals attribute to them (Myers 2011). The original study used interviews to collect data; 60 interviews were conducted with different organizational actors, and Grounded Theory was used to analyze the data. The interviewees were employed at three Portuguese companies and were at different hierarchical levels: peers HR managers, and employees from different organizational functions. These companies are mainly large companies and leaders in their fields of business and

Table 1 Classification of the data sample

Business clusters	Activity	Number of interviews by company
Portuguese	Industrial	25
Companies	Technology	25
	Commercial	26

were selected only companies with an HR department and an HR manager who is hierarchically and/or functionally subordinate to an administrator, general manager or superior general HR manager.

The theme of this study was not the objective of the global project, however, when it appeared spontaneously in these companies, it was understood by the relevance of the subject to develop and deepen it. As argued by Laperrière (2010), this study, focused on malpractices of HRM, constitutes a marginal case that gained central importance, by revealing a crucial aspect of the researched phenomenon.

The companies' classification, sectors of activity, and the number of interviews conducted are described in Table 1.

The interviewees covered a broad age range, from 19 to 82 years of age, and labor seniority, ranging from 3 months to 67 years, and were at different stages of their careers. The gender distribution of the interviewees was balanced across professional categories. Most of the interviews were carried out between 2007 and 2012. Given the changes in the financial and political conditions of the Portuguese context that occurred after the interviews were transcribed, additional interviews were conducted between 2013 and 2015 and confirmed the previous results.

In terms of data analysis, it should be noted that the domains created and categories are contextualized by reference to situations experienced or witnessed in their companies by the interviewees. The higher prevalence of reports and testimonies of this particular type of HRM activity is mentioned in three larger national companies by undifferentiated workers in the operational and administrative areas (12 interviewees), some line managers (6 interviewees), technical staff (6 interviewed), and some testimonies from peer directors of the HR director (36 interviewees).

4 Malpractices in HRM: Some Perceptions

Based on the interviewees' perception, it can be highlighted that exists a dark side within HRM reality. The perceptions about malpractices in HRM emerge from the interviews when the participants related effective situations. According to the analysis of the data, malpractices in HRM are associated with six dimensions: *perception of HRM;perception about the HR director and other managers;perception of determinants of the position of HR Director;perception of internal relationships and interpersonal interactions in the company,perception of HRMpractices.* This chapter is focused on the following dimensions: *perception of HRM;perception about the HR director and other managers; perception of determinants of the position of HR Director*.

4.1 First Dimension: Perception of HRM

In the dimension of *perception of HRM*, the results refer to one domain designed *basic principles of HRM* with two categories: relevance given to values and perception of coherence between rhetoric and practice, as presented in Table 2.

Regarding the *role of values* characteristic, there are 29 references, out of 60 possible, that state that although the values are considered very relevant, as they act as guidelines for behaviors, the reality is that they are perceived as something that:

> ... easily put aside. That is, people may have some purposes of following the values, but, more often than not, this aspect is no longer a priority, soon forgotten ... leaders should set an example but on the contrary. There are some here that are truly averse to following principles and coordinating their team following values ... (Technical Employee).

Regarding the *degree of perceived importance* characteristic, in 60 interviewees 33 references emerge in which the dominant perception is that "the values here in the company are often a dead letter" (Undifferentiated Operational Employee).

The second category that results from the data in a very prominent way has to do with the *perception of coherence between rhetoric and practice* in HRM. (42 refs). At this level, the data highlights one of HRM's chronically current issues, the gap between the discourse on the importance of people and the organizational reality. This gap exists mainly in the daily implementation of HRM practices and actions, embodied in the behaviors assumed by some of its protagonists. The characteristic underlying this category has to do with the *degree of consistency in HRM*, in which the following transcript exemplifies this perception and opinion: "In speech we are placed in the heart and in practice we are cannon fodder, unfortunately this is the truth" (Undifferentiated Operational Employee).

Table 2 First domain: basic principles of HRM

Categories	Characteristics	Perceptions
Relevance given to values	Role of values Degree of perceived importance	"Something that is easily put aside" "worthless piece of paper"
Perception of coherence between rhetoric and practice	Degree of consistency in HRM	gap between discourse and practical activities

4.2 Second Dimension: Perception About the HR Director and Other Managers

In the dimension designed by *perception about the HR director and other managers*, the results show three domains: *structural and hierarchical position of HRM*, *leadership of the diferent types of managers*, and *academic qualificationof the HR director*.

Regarding the first domain, there are some associated categories, as shown in Table 3.

The category linked to top management (41 refs) presents the characteristics of degree, nature, and type. In terms of degree, the perception is expressed not so much in relation to the greater or lesser distance between the HR director and the top managers of the company, but more in the quality of this type of relationship, resulting from the analysis of the data the balanced position that should exist in that relationship.

The opinion of a significant number of the interviewees (28 out of 60) can be summarized in the following quote, which shows the appropriate dimension of the relationship, but also what happens when the HR department does not function as a guarantor of impartiality and fair treatment:

> The HR director should have a lot of support from the top management ... I am talking in terms of what I think here at the company. In theory and in practice, I understand that a company must respect its workers and be consistently fair in the way they manage them and here the HR director must have an essential role. It is not, in practical terms, as I mentioned, what I actually see in this company. I believe that the link between top management and the HR director must be effectively close and supportive. But the principles and policies that link them must be in line with impartiality and justice, without jeopardizing the company's success, the quality of life of workers and their happiness, which in my view are not incompatible aspects. Now if the management says kill and the HR director, and his associates, say skinning, there is nothing good about this relationship, and we are the ones suffering because it's always the little guy who gets the raw deal, for many old wives' tales they tell or want to tell. I also tell you that sometimes management thinks and does one thing and at lower levels, for the sake of affirming their little power, they do it differently and it seems that the truth is not like oil, it takes forever before it reaches the top ... we died before. (Undifferentiated Administrative Employee)

Table 3 First domain: structural and hierarchical position of HRM

Categories	Characteristics	Perceptions
Relationship with top management	Degree	Level of proximity to top management
	Nature	Excessive bonding to top management Instrumental use of the relationship with top management Lack of courage in making business decisions Convenience positioning
	Type	Instrumental opportunity Partial orientation

It also follows from the data and the previous quote that, in terms of HRM performance, the interviewees, although they manage to differentiate the performance (or inactivity) of the HR director, tend to see decisions in HR matters associated with other actors with management responsibilities, whether they are peer directors of the HR director or line managers. The data reveal that the HR director's inadequate perception "creditor of certain abusive behaviors that should not occur". (Undifferentiated Administrative Employee)

In terms of the nature of the positioning, the perception is that

> … there is a tendency for the HR director not to face many of the situations head-on, I mean clearly the most controversial, unpopular as you want to call it and safeguard himself in "them", like: «it was the management that decided, it has to be like this ». In my opinion, this type of posture discredits the position and the function, puts himself in a position of minority and as a rubber stamp, without a voice, crouching in front of the Administration. (Suj. Quadro Técnico) (31 refs out of 60)

Data analysis reveals that regarding the nature of positioning, the most common perceptions are excessive bonding to top management, instrumental use of the relationship with top management, and lack of courage in making business decisions, *"so as not to have to face the decisions that have to be taken"* (Undifferentiated Operational Employee) or become strongly associated with those decisions. These perceptions contribute to the discrediting and devaluation of the professional role of the HR director and, in turn, this deletion favors the existence of bad HR management behaviors, which tend to pass unscathed.

The transversality of HRM between the different management positions is also revealed here, as it also emerges from the data that "when it comes to bad news, the boss tells us to talk to HR" or "these situations are the responsibility of the HR management, I don't even agree, but they in HR want it that way". Thus, there is a perception of not taking responsibility for people or teams when "things are less good" (Technical Employee).

In the second domain, *leadership of the different types of managers*, the categories that emerge were designated by *leadership style* (47 refs) and *leadership exercise* (48 refs), as shown in Table 4. The similarity of references reveals an identical perception in terms of the listed categories.

Regarding the leadership style, data appear as shown in the following quote:

Table 4 Second domain: leadership of the different types of managers

Categories	Characteristics	Perceptions
Leadership style	Assigned components and their designations	Passivity and fade away Abuse of the leadership position Abuse and discrimination
Leadership exercise	Form of leadership exercise Determinants leadership exercise	"Non-leadership" and/or "Schizophrenic leadership" Insecurity, ignorance, and lack of competence

There is a book by Mintzberg that, if I'm not mistaken, the title is «Less MBA's, More Managers» and what I see, and which you will find paradoxical in view of the position I have and the fact that I let go and don't do nothing… I acknowledge that in fact I do nothing because I am in a delicate moment of my life here at the company, as they are negotiating with me to terminate my contract by mutual agreement, and it is a very delicate and complex situation… but the truth is that, and believe that it is not because I am in the situation that I am in that I tell you this, it is difficult for me to see certain actions taken by some managers, who fell in favor of the gods and do and undo at their own pleasure, destroy much of the company's assets in terms of credibility, seriousness, image; they don't bother to mess up people's lives to the point of making it hell. The number of casualties confirms this situation, there are things that are not even worth talking about, very serious in terms of health (tears)..., and they still have plenty of time. I speak of heads who manage their areas and their people more in the style of animals than people. In my view, the company was filled with MBAs, boys and girls who have been trapping and literally cleaning up the existing critical sense, we are creating a set of management dictators… I have nothing against the MBAs, I have a son making one and I hope the title doesn't go to his head, but I don't put my hands on the fire anymore. In fact, the truth is that in terms of leadership everything is said and if only they were managers ..., but their attitudes and behaviors are unharmed and, of course, the HR director cannot do well in this photo because he often appears in it all smiling. (Manager Peer)

There are 26 interviewees who express ideas similar to those in the transcript. Thus, leaders perceived as being based on passivity and fade away are translated into the word "disappearance" of the director in the face of different situations. In contrast, 18 of the 26 interviewees also express the idea of managers who stand out for the abusive use of leadership positions translated into behavior of arrogance and discrimination.

Regarding the *leadership exercise* category, two characteristics of the data analysis emerge very explicitly: the *form* and the *determinants of the leadership exercise.* Regarding the form, the perception that comes up most often is the arrogance in the exercise of functions, which leads some interviewees to speak of "non-leadership" (Undifferentiated Employee). There are also references such as "To terrorize and discriminate is not to lead" (Undifferentiated Employee), and "Dealing with people with insults and disrespecting and trampling is not for a leader, it is someone who thinks 'I want, I can and I command'" (Qualified Technical Employee). Another aspect regarding the exercise of leadership is the consideration of a schizophrenic way of leading. This idea can be translated into the following quote: *"When you order something and its opposite, when you ask for something and it is done exactly as it was requested and then it is simply destroyed, in the literal sense of the term, kicking or verbally, you cannot be in perfect judgment, this is crazy…"* (Undifferentiated Administrative Employee) (11 refs out of 60). Some of the interviewees use besides "schizophrenic" the word "autstic" (7 refs).

Insecurity, ignorance, and lack of competence emerge in the determinants, which can condition and lead to an overbearing exercise of the leadership of some managers and which is illustrated in the following transcript:

… Leadership is something that cannot described, because what exists is a tyrannical way of exercising power and, in my opinion and I certainly will not be the only one, this power is exercised in this way to compensate for insecurities, ignorance, unpreparedness and even

Table 5 Third domain: academic qualification of the HR director

Categories	Characteristics	Perceptions
Type of school education	Degree and area of training	Exercise of the positions markedly ideological and biased prejudice, result of the academic background
Consequences of different school backgrounds	Degree of suitability	Blurred and misadjusted academic perspective of reality

fear. I refer to managers who do what they want and who in HRM are a disaster and the greater the disaster the greater the tyranny… (Undifferentiated Operational Employee) (31 refs).

In the *academic qualificationof the HR director* domain, the categories that arise have to do with the *type of school education* and the main *consequences of different school backgrounds*, as in Table 5.

At this level, the analysis of the data reveals that some interviewees, despite assuming that it is important for HR directors to have higher education, question that same higher education through the opinions they express:

…To what extent does the higher education acquired allow future professionals in this area to have a comprehensive, integrative and as inclusive view as possible about people management issues? That is, the extent to which higher education is structured so that future leaders have a management conscience in which all people count or, on the contrary, higher education is structured to train professionals with a markedly ideological and prejudiced matrix, focused only on the interests of the company. (Technical Employee)

Regarding the *consequences of different school backgrounds* category, a characteristic called *degree of suitability* of higher education to business reality emerges. The idea transmitted by 26 interviewees is that school education "is out of touch with reality", "is too dark when students action scenarios are already very black" (Undifferentiated Operational Employees).

In terms of data analysis, the question that arises in this domain and goes through the two categories mentioned above is related not so much to the type, area and consequences of higher education, but to the way in which the knowledge resulting from that higher education is properly implemented in practice. From the interviews carried out in the three national companies that have served to illustrate and analyze ways of exercising HRM, the following transcript is highlighted regarding the level of desirable in terms of higher education:

It is essential that the HR director, regardless of the titles and academic degrees he has, has the ability to see, hear and feel. Be alert. Knowing how to listen in the midst of noise and seeing in the middle of smoke is not taught in colleges and courses. Unfortunately, not being able to understand the symptoms of arrogant behaviors is also not taught, but it should be, as it would certainly avoid certain behaviors that are often destructive of the value created and would encourage behaviors that would be a stimulus for the development of companies and for the involvement of all those who work at the company. (Technical Employee) (34 refs out of 60)

According to the data, it appears that this lack of adequacy of school education to practical reality has the following main consequences: enhancing the partiality of the HR director and HRM; foster corporate hypocrisy by employees who have leadership responsibilities and other organizational actors; and, increase business distrust, as in the following transcripts:

> I can't imagine how many times I feel that they treat me with hypocrisy and I put on a smile that no matter how much I want to hide it is always yellow ... it is a case of saying that hypocrisy with hypocrisy pays off. When I feel like I give everything and my colleagues too and we are treated cynically and hypocritically, by hypocritical people, whom I neither recognize competence nor values, but just because they have the king in their belly they think they can treat us like doormats, this it is not good for my health or for the company… (Undifferentiated Operational Employee) (17 refs)

> … Training is very important, but more important in my opinion is what we do with it and from it. In some cases, what I see at the highest level, and I am talking about people with management positions, to make it clear is not the HR director, but other directors who do, undo, redo at their pleasure. People who lie, harm the company, create a climate of fear such that in management meetings no one speaks. … sometimes I think their eyes will pop out, such is the verbal, psychological and even physical violence, kicking the poor furniture that it is impossible for anyone to say anything. I already talked to whoever had to speak and I think they left me with the grenade in my hands, without a peg and captured. I ask: higher education for what? The company where I started fit in the reception and we were half a dozen, but it was classy. This has already been, but has been living in this environment, and then I no longer understand the position, power or attention of the HR director. Doesn't he see or doesn't want to see!? I just want the time to get out of here (Manager Peer).

4.3 *Third Dimension: Perception of Determinants of the Position of HR Director*

There is one domain that results from the analysis of the interviews in this dimension, which has to do with the *perception of the positionof the HR director*. In the Portuguese companies analyzed, there was a greater number of references by different actors to situations characterized as "… bad or very bad HRM" (Undifferentiated Operational Employee). These perceptions strongly affect the perception of the HR director and other actors with hierarchical and functional responsibilities in this area of management.

The perception of the position of the HR director emerges as transversal to the three participating companies. This situation can be illustrated by the following transcript:

> What I can tell you is that it is almost impossible not to find companies where at one time or another in their existence they have not had a moment of rotten peace generalized or even circumscribed to a certain area. Of course, when I talk about rotten peace, I mean the climate that is established between workers of the same company in the performance of their duties and as such it has a lot to do with human interactions and consequently with HRM. In my opinion, the HR director should act as an element of detection and search for solutions that aim to eliminate this type of environment, however the reality that I know shows me that sometimes the greatest difficulty is in being able to diagnose this climate of rotten peace. The image is a little harsh, but I compare the rotten peace to a cancer that can often progress

Table 6 First domain: perception of the position of the HR director

Categories	Characteristics	Perceptions
Characteristics of the HR Director position	Requirements for the position	Deficit of social skills
	Supremacy of individual logic over group and company logic	Narrative contrary to "flying the flag of the company"
	Behaviors of the HR director resulting from the exercise of the position	Obsessive and aggressive verbal behaviors Breach of trust in the direction and director of HR Discredited in many HR practices Lack of empathy and business solidarity Abuses of power, sexual and moral harassment and psychological terrorism
Limitations arising from the exercise of the position	Suppression of critical thinking	"You are here to work and not to think"

> silently, without symptoms worthy of highlight and when it manifests itself it is already fatal. (Manager Peer)

In the domain of this third dimension, perception of the position of the HR director, two categories are identified, each with different characteristics, as shown in Table 6.

Regarding the first category designated as *characteristics of the HR director position* (34 refs out of 60) and its first characteristic—*requirements for the position*—in the analysis carried out in these three companies, the perception goes beyond the required profile. What emerges from the data in relation to the requirements, also designated by the interviewees by competencies, is in accordance with the following quote:

> … the existence of a deficit in social skills, namely in terms of the quality of interaction management, communication management and, essentially, expectations management. People are concerned about their lives and about their benefits and their rights. Here many of those who say that workers only know how to talk about rights, you should see and hear them in private, because in fact they only talk about their rights, benefits and incentives. We talk about overtime that they don't want to pay us, and at the top level what we hear, in a blatant and scandalous way, is the cylinder capacity of the car assigned by the company, the limit and the color of the credit card, the size of the office, the quality of the laptop and the capabilities and design of the mobile phone. And, of course, the HR director puts himself in the same perspective and takes care of his life. Of course, these types of situations are unfair, but it says a lot about the management mindset. Cows are always skinny for them, but chubby and dairy for the usual ones. At least there should be a minimum of decorum and discretion. In relation to the HR director, he follows the same gauge as the others, so anyone who doesn't want to be a wolf should not wear his skin. (Undifferentiated Operational Employee, finalist of a degree in HRM)

The second characteristic—supremacy of individual logic over group and company logic—is the opposite of the "flying the flag of the company" narrative, and can be substantiated in the expression by an undifferentiated administrative employee:

> … flying the flag of the company at all time is nothing when compared to what is required of us by certain company leaders. They pay us for our work, but most of the time they don't pay us for the extra hours we work, so flying the flag of the company is one thing, leaving the skin on and getting screaming and bad manners isn't a fair price, don't you The department and the HR director don't see, don't hear, don't know? However, if one of us says he can't work overtime, he is immediately called or threatened with the HR director. Of course, I even admit that he may not know and that his position and person are used to pressure and frighten. There is still a lot of mentality that fear is an excellent way to motivate ..., but the HR director should know how to identify these situations and act, in fact, he gets the fame! (14 refs out of 15 undifferentiated administrative and operational employees).

Regarding the third characteristic–*behaviors of the HR director resulting from the exercise of the position* (34 refs)—the perception that exists, which according to some interviewees results from what they heard from the HR director himself in different contexts, is the existence of obsessive and aggressive verbal behaviors to achieve results at any cost, as transcribed below

> This dictatorship of results at any price leads to disrespect for personal and professional dignity, with the increase in dismissals. The fact that there is more work force available leads, in terms of recruitment and selection practices, to the use of cheap and unskilled labor and situations of contractual precariousness, which has everything to go wrong. … In fact, people already work in some companies that I know without receiving any remuneration and I am not thinking about overdue wages, they simply do not receive, they are young and the parents pay food, transportation, for them to work. … The famous professional internships… and I believe this is going to get worse, I would like to be mistaken, but as I told you, this is taking a course that has everything to end badly. (Technical Employee)

In terms of perceived behaviors, there is also: *breach of trust in the direction and director of HR* (23 refs); *discredited in many HR practices* (29 refs) seen as ways of homogenizing workers and putting pressure on them; *lack of empathy and business solidarity* (37 refs), and finally, the existence of

> … abuses of power, sexual and moral harassment and psychological terrorism that I never thought I would see on the part of managers appointed with pomp and circumstance by the company's senior officials. … now you think I say this out there, of course not, even if I had one another guaranteed job, because if those who can do it don't do it, am I going to be the savior of the country?! But there is still worse than that in other companies and nothing ever happens, the victims are the ones who end up being fired. (Undifferentiated Operational Employee).

As for the second category in this domain—*limitations arising from the exercise of the position* (18 refs)—it has a characteristic: suppression of critical thinking, expressed in the following quote:

> It is our daily bread to hear the phrase "you are here to work and you are not here to think", of course this is an excellent way to motivate, to involve, to engage people in the company's goals. The most cynical of all this is that after forbidding us to think, they hire a company

> or someone to come and give us self-motivation training. Or the managers will all learn to motivate their teams, or consultancy companies will come and interview us to find out what we think. Extraordinary, not even Tarantino remembered to conceive and film this type of dialogue. (Undifferentiated Operational Employee)

The data reveal that more important than the position and the role associated with the HR director is the way in which this position and associated role is exercised, and also, the way in which it can directly or indirectly influence the exercise of practices and HR management actions by other organizational actors with management positions.

5 Conclusion

The identification of malpractices of HRM may and should work not as way to only bring to light what is wrong, but as a way to learn how to improve and innovate in HRM practices, such as coaching as a way to enhance psychological safety and drive organizational effectiveness (Graen et al. 2020). It is fundamental that HR managers work as team leaders and potenciate the creation of friendly and stimulating environments. This study focused on understanding how the way in which the position of HR Director is exercised, and its influences on the performance of other management positions, can give rise to poor HRM strategies, behaviors, and situations. The study calls for a deep understanding of those malpractices in order to allow the flourishing of best practices in HRM, in particular for stimulating creative and innovative teams, such as the creation of a team coaching staff in organizations.

The position of HR director is highlighted by the interviewees taking into account what the HR director does, why he does it and what he does it for, but in many of the transcripts presented, it is marked by its deletion, disappearance, lack of management assertiveness. This results in a perception of complicity, albeit by inaction, in relation to certain situations and behaviors of poor HRM.

The cross-cutting of HRM across different hierarchical levels leads, according to the data, to the fact that HR responsibilities may be assumed by managers with little technical knowledge and little behavioral skills, in an area of management where, according to many of the interviewees, it will be necessary to have well-trained people at all levels.

The little respect for the values and the demagogy of the discourse about the importance of people as the most valuable capital of an organization leads, in face of what is practiced on a daily basis, to a diminishing of HRM, the HR department, managers and the organization. These circumstances create a favorable environment for "I want, I can and I command" and for the destructive powers in which the only prevailing logic is that of personal interest (Hershcovis 2011; Krasikova et al. 2013; Tepper et al. 2017).

Aggressive, offensive, discriminatory, abusive, destructive, persecutory behaviors, which disrespect the citizenship and dignity of the worker and the person, translate into anxiety, stress, various diseases and disorders, absenteeism and high turnover. These behaviors and their implications associated with the impunity of those responsible, generate perceptions that in turn discredit the company and make it a place to avoid or leave the company as soon as possible (Hershcovis 2011; Krasikova et al. 2013; Morten and Einarsen 2018; Tepper et al. 2017). In addition, precarious work can generate the feeling that workers are "flying the flag of the company", because they stay and work there. However, the reality may be, judging by the data obtained, that people return the next day because they have no other work alternatives and continue to have commitments and bills to pay at the end of each month (Workplace Bullying Institute 2014).

The fact that the companies analyzed in this study have an HR department and an HR director necessarily raises expectations regarding this area of management, especially since some of the interviewees came from previous work experiences where these policies and practices were not formalized. These expectations were reflected in statements about the real importance of having an "HR management with pomp and circumstance", assuming itself as the guarantor of good practices in the company and ensuring that workers and managers, having different goals and expectations, can contribute to the sustained development of the organization as a whole (Guest 2017; Hershcovis 2011; Krasikova et al. 2013; Tepper et al. 2017).

The greatest limitation of this work results from the fact that the sample is limited and above all because this theme was not foreseen in the original research project. It resulted from the spontaneity of the interviewees. Accordingly, a proposal for future research is to deepen the data obtained and have this theme as the central focus of an investigation in the Portuguese reality and with different types of organizations involved.

References

Aborde de Chatillon, E., & Richard, D. (2015). Du Sens, du Lien, de l'Activité et de Confort (SLAC): Proposition pour une Modélisation des Conditions du Bien-être au Travail par le SLAC. *Revue Française de Gestion, 4*(249), 53–71.

Astrauskaitė, M., Perminas, A., & Kern, R. M. (2010). Sickness, colleagues' harassment in teachers' work and emotional exhaustion. *Medicina, 46*(9), 628–634.

Atwater, L. E., Waldman, D. A., Atwater, D., & Cartier, P. (2000). An upward feedback field experiment: Supervisors' cynicism, reactions, and commitment to subordinates. *Personnel Psychology, 53,* 275–293.

Bernotaute, L., & Malinauskiene, V. (2017). Workplace bullying and mental health among teachers in relation to psychosocial job characteristics and burnout. *International Journal of Occupational Medicine and Environmental Health, 30*(4), 629–640.

Burton, J. P., & Hoobler, J. M. (2011). Aggressive reactions to abusive supervision: The role of interactional justice and narcissism. *Scandinavian Journal of Psychology, 52*(4), 389–398.

Burton, J. P., Taylor, S. G., & Barber, L. K. (2014). Understanding internal, external, and relational attributions for abusive supervi-sion. *Journal of Organizational Behavior, 35*(6), 871–891.

Crawshaw, L. (2009). Workplace bullying? Mobbing? Harassment? Distraction by a thousand definitions. *Consulting Psychology Journal: Practice and Research, 61*(3), 263–267.

Daniels., K., Watson, D., & Gedikli, C. (2017). Well-being and the social environment of work: A systematic review of intervention studies. *International Journal of Environmental Research and Public Health, 14*(918), 1–16.
Davis, W. D., & Gardner, W. L. (2004). Perceptions of politics and organizational cynicism: An attributional and leader–member exchange perspective. *The Leadership Quarterly, 15,* 439–465.
Einarsen, S. V., Hoel, H., Zapf, D., & Cooper, C. (2019). *Bullying and harassment in the workplace: Theory, research and practice (3a* (Edição). Florida, EUA: CRC Press, Taylor and Francis Group.
Einarsen, S., Hoel, H., Zapf, D., & Cooper, C. (2003). The concept of cullying at work: The European tradition. In S. Einarsen, H. Hoel, D. Zapf, & C. Cooper (Eds.), *Bullying and emotional abuse in the workplase* (pp. 3–30). London: Taylor & Frances.
Eurofund, EU-OSHA. (2014). Psychosocial Risks in Europe: Prevalence and Strategies for Prevention. https://osha.europa.eu/en/publications/psychosocial-risks-europe-prevalence-and-strategies-prevention/view. Accessed 14 Aug 2020.
Fahie, D., & Devine, D. (2014). The impact of workplace bullying on primary school teachers and principals. *Scandinavian Journal of Education Research, 58*(2), 235–252.
Graen, G., Canedo, J. C., & Grace, M. (2019). Team coaching can enhance psychological safety and drive organizational effectiveness. *Organizational Dynamics, 49*(2), 1–6.
Guest, D. (2017). Human resource management and employee well-being: Towards a new analytic framework. *Human Resource Mangement Journal, 27*(1), 22–38.
Hershcovis, M. S. (2011). "Incivility, social undermining, bullying… oh my!": A call to reconcile constructs within workplace aggression research. *Journal of Organizational Behavior, 32,* 499–519.
Katz, D., & Kahn, R. L. (1978). *The social psychology of organizations* (2nd ed.). New York: Wiley.
Khoo, S. B. (2010). Academic mobbing: Hidden health hazard at workplace. *Malaysian Family Physician, 5*(2), 61–67.
Kowalski, T. H. P., & Loretto, W. (2017). Well-being and HRM in the changing workplace. *the International Journal of Human Resource Management, 28*(16), 2229–2255.
Krasikova, D. V., Green, S. G., & LeBreton, J. M. (2013). Destructive leadership: A theoretical review, integration, and future research agenda. *Journal of Management, 39,* 1308–1338.
Laperrière, A. (2010). A teorização enraizada (grounded theory): Procedimento analítico e comparação com outras abordagens similares. In J. Poupart, J.-P. Deslauriers, L. Groulx, A. Laperrière, R. Mayer & A. Pires (Eds), *A pesquisa qualitativa: Enfoques epistemológicos e metodológicos* (pp. 353–385), Tradução de Ana Cristina Nasser (2ª Ed.). Petrópolis, RJ: Vozes.
Legge, K. (1995). *Human resources management: Rhetorics and realities.* London: The Macmillan Press.
Martinko, M. J., Harvey, P., Brees, J. R., & Mackey, J. (2013). A review of abusive supervision research. *Journal of Organizational Behavior, 34,* 120–137.
Morten, M. B., & Einarsen, S. V. (2018). What we know, what we do not know, and what we should and could have known about workplace bullying: An overview of the literature and agenda for future research. *Aggression and Violent Behavior, 42,* 71–83.
Myers, M. D. (2011). *Qualitative research in business and management.* London: Sage.
Teixeira, R., Munck, L., & Reis, M. (2011). Assédio moral nas oganizações: Perceção dos gestores de pessoas sobre os danos e políticas de enfrentamento. *Revista Gestão Organizacional, 4*(1), 31–48.
Tepper, B. J. (2000). Consequences of abusive supervision. *Academy of Management Journal, 43,* 178–190.
Tepper, B. J. (2007). Abusive supervision in work organizations: Review synthesis, and research agenda. *Journal of Management, 33,* 261–289.
Tepper, B. J., Duffy, M. K., Henle, C. A., & Lambert, L. S. (2006). Procedural injustice, victim precipitation, and abusive supervision. *Personnel Psychology, 59,* 101–123.
Tepper, B. J., Simon, L., & Park, H. M. (2017). Abusive supervision. *Annual Review of Organizational Psychology and Organizational Behavior, 4*(1), 123–152.
Urquhart, C. (2013). *Grounded theory for qualitative research: A practical guide.* London: Sage.

Valle, M., & Perrewe, P. L. (2000). Do politics relate to political behaviors? Tests of an implicit assumption and expanded model. *Human Relations, 53*(3), 359–386.

Delfina Gomes acknowledges that this study was conducted at the Research Center in Political Science (UIDB/CPO/00758/2020), University of Minho/University of Évora, and was supported by the Portuguese Foundation for Science and Technology and the Portuguese Ministry of Education and Science through national funds.

Correction to: Coaching for Managers and Engineers

Carolina Machado and J. Paulo Davim

Correction to: C. Machado and J. P. Davim (eds.), *Coaching for Managers and Engineers*, Management and Industrial Engineering, https://doi.org/10.1007/978-3-030-71105-4

In the original version of the book, the following belated corrections have been incorporated in Chapters "From Theory to Practice—in Search of Theoretical Approaches Leading to Informed Coaching Practices" and "Coaching for All—New Approaches for Future Challenges" belated corrections has been incorporated as in below:

The affiliation "Instituto Superior Manuel Teixeira Gomes, Universidade Lusófona, Lisbon, Portugal" of author "Carla Gomes da Costa" has been changed to "Instituto Superior Manuel Teixeira Gomes, Lisbon, Portugal".

The updated version of these chapters can be found at
https://doi.org/10.1007/978-3-030-71105-4_1
https://doi.org/10.1007/978-3-030-71105-4_4

C. Machado and J. P. Davim (eds.), *Coaching for Managers and Engineers*,
Management and Industrial Engineering,
https://doi.org/10.1007/978-3-030-71105-4_7

Index

C. Machado and J. P. Davim (eds.), *Coaching for Managers and Engineers*, Management and Industrial Engineering,
https://doi.org/10.1007/978-3-030-71105-4

Printed in the United States
by Baker & Taylor Publisher Services